Using Stata for Quantitative Analysis

Kyle C. Longest
Furman University

Los Angeles | London | New Delhi
Singapore | Washington DC

Los Angeles | London | New Delhi
Singapore | Washington DC

FOR INFORMATION:

SAGE Publications, Inc.
2455 Teller Road
Thousand Oaks, California 91320
E-mail: order@sagepub.com

SAGE Publications Ltd.
1 Oliver's Yard
55 City Road
London EC1Y 1SP
United Kingdom

SAGE Publications India Pvt. Ltd.
B 1/I 1 Mohan Cooperative Industrial Area
Mathura Road, New Delhi 110 044
India

SAGE Publications Asia-Pacific Pte. Ltd.
33 Pekin Street #02-01
Far East Square
Singapore 048763

Executive Editor: Jerry Westby
Production Editor: Brittany Bauhaus
Copy Editor: QuADS (P) Ltd.
Typesetter: C&M Digitals (P) Ltd.
Proofreader: Eleni-Maria Georgiou
Cover Designer: Anupama Krishnan
Marketing Manager: Erica DeLuca
Permissions Editor: Karen Ehrmann

Printed in the United States of America

Library of Congress Cataloging-in-Publication Data

Longest, Kyle C.

Using Stata for quantitative analysis/
Kyle C. Longest.

p. cm.
Includes bibliographical references and index.

ISBN 978-1-4129-9711-9 (pbk.)

1. Stata. 2. Social sciences—Graphic methods—Computer programs. 3. Social sciences—Statistical methods—Computer programs. I. Title.

HA32.L66 2012
005.5′5—dc23 2011041851

Certified Chain of Custody
Promoting Sustainable Forestry
www.sfiprogram.org
SFI-01268

SFI label applies to text stock

12 13 14 15 10 9 8 7 6 5 4 3 2

Brief Contents

Detailed Contents

Preface

Motivation and Purpose

The motivation for this book, as I assume is true for most, came from a series of personal experiences. First, as a graduate student, I remember literally laying awake at night dreading the idea of using a computer program to conduct statistical analyses. The first statistics course I took required Stata to complete the assignments and the final research project. This necessity was so overwhelming at the time, in part, because there did not seem to be any straightforward, concise texts explaining the basics of Stata. Over my time in graduate school, I came to be very familiar with Stata, even to the point that I developed a serious passion for both learning Stata and teaching it to students who were facing the same fears I once did. In a somewhat mirrored experience, I was hoping to use Stata as a significant portion of the classroom experience and requirements when I first began teaching a course on Quantitative Analysis. I soon realized that there still was not a manageable introductory text on the use of Stata for quantitative research.[1] Thus, I sought to contribute to filling this void by providing a straightforward, applied introduction to using Stata.

This book will be most beneficial to readers who are novices when it comes to Stata and are at least in the early stages of learning strategies for conducting quantitative analysis. It does assume that the reader has a working knowledge of basic statistical techniques and terminology. The organization and coverage of the book is guided by the content and ordering of topics found in most introductory social statistics textbooks. In this manner, it can serve as an excellent companion, either for a class or self-learner, to such a textbook.

[1]Assuredly, there are several very good and effective texts on learning Stata. Virtually all of these, however, are aimed at experienced users or are so detailed and long that they are not helpful for a typical classroom in which teaching Stata is not the primary purpose.

To be clear, this book should not be used to learn statistics or quantitative analysis. Some basic assumptions and explanations are provided, but these should not be used in place of a more thorough coverage of each of the analytic strategies. The statistical grounding for this book is based primarily on Frankfort-Nachmias and Leon-Guerrero's (2009) *Social Statistics for a Diverse Society*. The definitions and interpretations of the specific measures and tests are based on those presented in this text. Of course, any inaccuracies or mistakes are solely mine.

Also, this book does not attempt to cover every aspect of each Stata command that is introduced. More experienced users undoubtedly know shortcuts or alternative methods for the techniques that are presented. But the given description has been geared to introduce complete novice users to Stata. This targeted audience requires that the explanation starts with the basics before jumping into the advanced features. The presented commands and procedures are discussed because they are the most simplified strategies that effectively accomplish the pertinent goals.

About the National Study of Youth and Religion

The data for this book come from the National Study of Youth and Religion (NSYR). The NSYR is a longitudinal, nationally representative telephone survey of U.S. young adults. There are three waves of data, all of which are publically available.

The variables that are used in the examples throughout this book come from the most recent follow-up survey of 2,532 young adults completed in the fall of 2007. At the time of this survey, the respondents were all between the ages of 18 and 24. Each respondent completed a computer-assisted telephone interviewing (CATI) survey that lasted approximately an hour. This data set covers a broad array of topics, making it possible, across examples, to use variables pertinent to several disciplines. For example, it contains several standard self-esteem measures of interest to psychologists, a wide array of questions on religion useful for sociologists, numerous questions on finances (e.g., debt) applicable to economics, and measures of substance use behaviors that would be pertinent to social work or health researchers. The full data set and documentation can be downloaded from the Association of Religion Data Archives (http://www.thearda.com/Archive/Files/Descriptions/NSYRW3.asp).

The first wave of the survey sampled 3,290 U.S. English- and Spanish-speaking teenagers, ages 13 to 17. The sampling and survey were conducted from July 2002 to August 2003 using random-digit-dialing, drawing on a sample of randomly generated telephone numbers representative of all noncellular phone numbers in the United States. The overall response rate of 57% for the first survey is lower than desired, but it is similar to other current nationally based surveys using similar methodologies. Further comparisons of the NSYR data with 2002 U.S. Census data on households and with nationally representative surveys of adolescents—such as Monitoring the Future, the National Household Education Survey, and the National Longitudinal Study of Adolescent Health—confirm that the NSYR provides a nationally representative sample of U.S. teenagers aged 13 to 17 years and their parents without identifiable sampling or nonresponse biases (for details, see Smith & Denton, 2005). The follow-up sample that is used in the data sets comes from this initial sample of 3,290 teens. To obtain more information regarding the technical details and documentation of the NSYR, please visit http://www.youthandreligion.org/.

A Note on Versions

All the commands and examples for this book were produced using Stata 12.0 for Windows. The primary commands and options are similar for older versions, dating back until at least Stata 9. There were, however, a few changes between Stata 11 and Stata 12. Most of these changes do not affect the actual functionality but rather deal with convenience and appearance. In fact, most of the substantive differences that the new users would encounter fall under the topics covered in Chapter 1.

Due to the very recent release of Stata 12 (July 2011), many readers may still be using Stata 11 or even Stata 10. To address this potential challenge, this book includes two versions of the introductory material (i.e., Getting to Know Stata). The vast majority of the material in both versions is extremely similar, but both were included to prevent any confusion over the small dissimilarities. For users of Stata 12, please start with Chapter 1: Getting to Know Stata 12. For users of Stata 11 (or older), please start with Appendix: Getting to Know Stata 11, and then rejoin the book at Chapter 2. From that point on, all of the commands and strategies are equivalent across versions (although the appearance of the screenshots may be slightly different).

The vast majority of the commands presented are similar for Stata for Mac as well. The appearance and wording of some icons as well as the pathways for the point-and-click menus may be slightly different for a Mac operating system.

A Note on Notation

Certain text in this book will be presented in a slightly different font. Generally, anything that you enter into or that comes out of Stata will be denoted with the `typewriter` (i.e., Courier New) font. This font will be used to indicate variable names in a particular data set, such as `gender` or `ids`. It will also be used to show the display from the Stata Results window (if the actual screen shot is not shown).

This font will be used to denote a command that is entered into the Command window to perform a given operation. Additionally, if these commands are presented by themselves within a sentence, they will be set apart by a dash pre and post (e.g., `-replace-`) so that they are not confused with a variable name.

The majority of this book discusses the syntax command interface (i.e., the Command window) aspect of Stata. But there will be times when the menu, point-and-click interface is described. Menus (e.g., File), clickable buttons (e.g., OK), or keys on the keyboard (e.g., Enter) will be denoted with the Arial font.

Finally, Stata is a case-sensitive program, meaning that all commands and variable names must be typed exactly as they are shown. For the purposes of this book, this sensitivity means that at times the capitalization may not follow typical grammatical conventions. For example, if a variable name starts a sentence and that variable name is lowercase, then that sentence will start with a lowercase letter.

References

Frankfort-Nachmias, C., & Leon-Guerrero, A. (2009). *Social statistics for a diverse society* (5th ed.). Thousand Oaks, CA: Pine Forge Press.

Smith, C., & Denton, M. L. (2005). *Soul searching: The religious and spiritual lives of American teenagers.* New York, NY: Oxford University Press.

Acknowledgments

The author and SAGE gratefully acknowledge the contributions of the following reviewers:

Karen Y. Holmes, *Norfolk State University, Norfolk*

Sean Kelly, *University of Notre Dame*

David Peterson, *Iowa State University, Ames*

Raymond Sanchez Mayers, *Rutgers University, New Brunswick*

PART I

Foundations for Working With Stata

1

Getting to Know Stata 12

For many people, learning any new computer software can be an anxiety-producing task. When that computer program involves statistics, the stress level generally increases exponentially. If you have similar feelings as you begin your journey into becoming a Stata user, do not fear, you are not alone. This book is designed with this apprehension in mind. One of the primary goals of this book is to help alleviate, or at least minimize, this anxiety as we move toward becoming an effective and proficient Stata user. Keep in mind that at one time you may have had similar feelings about using e-mail or the Internet, and just as many people now feel extremely comfortable using these programs, by the end of this book you will have a similar grasp of and comfort with Stata.

Before diving into all the details of using Stata, it is important to have an understanding of its various components. This chapter will serve as an introduction to the basic building blocks of the Stata program. Each of these aspects will be covered in much more detail throughout the book, but this chapter provides an overview of the basic functionality of the Stata program. The second section of the chapter explains how data are opened, imported, and entered.

What You See[1]

When you open Stata, by double clicking on the Stata icon, for the first time, you will see the following screen:

[1]If you are using Stata 11 (or Stata 10), please use Appendix: Getting to Know Stata 11 instead of this first chapter. All of the same features are covered, but Stata 12 has a slightly different appearance from these previous versions, which may make matching up what you see in the text and on your screen a bit confusing. Starting from Chapter 2, the vast majority of operations and commands are similar across versions. And the text specifically notes any particular features that are different for previous versions.

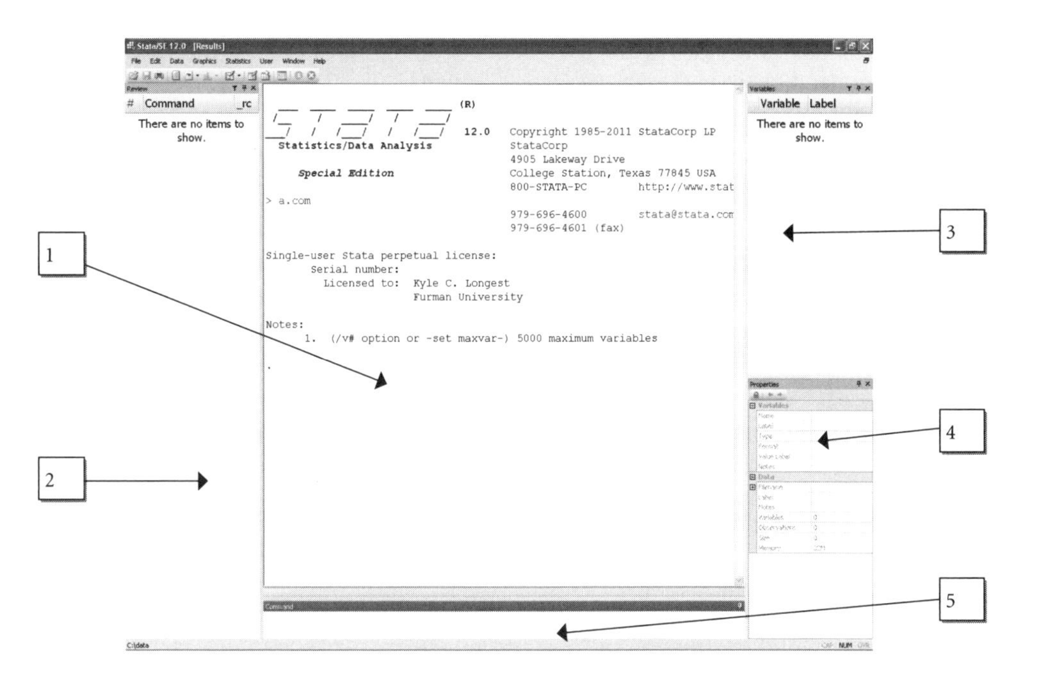

There are five different windows on the screen.[2]

1. Results Window: The Results window is where everything that Stata "does" will be displayed. Anytime Stata executes some operation, it will display that operation and its results in this window. These results, however, are not automatically saved. How to save these results is covered in the Data Management: Saving Results section of Chapter 3.

2. Review Window: The Review window contains a running history of all the operations that have been performed during the current session of Stata. Whenever you enter and execute a command, it will appear both in the Results window and in the Review window. The most useful aspect of the Review window is that it can be used as a shortcut to work with a previously executed command. When you click on a command in the Review window, that

[2]This layout is what you would see if Stata was opened "right out of the box." If you are working on a shared computer (or over a network), there is a chance that these windows have been moved, resized, or even deleted by another user, making what you see slightly different from the screenshot presented. If any of these windows are missing, you can click on the Windows tab and click on the desired window. You can also move these windows by simply clicking on them with your mouse and dragging them to the desired location.

command will appear in the Command window, from which you can alter the command or simply rerun the same command.

3. Variables Window: When you open a data file in Stata, the variables contained in that data set will be listed in the Variables window. This window can be used to scroll through and see all the variables that are contained in the active data. Whenever you click on a variable name listed in the Variables window, several characteristics of that variable are displayed in the Properties window. If you place your cursor over a variable, a small arrow will appear. By clicking on that arrow, the variable name will automatically appear in the Command window. This window also lists the variable "Label," which presents more detailed information about the variable. Labels are discussed in more detail in the Data Management: Working With Labels section of Chapter 3.

4. Properties Window: The Properties window provides details about the data set that is currently being used and any variable that has been selected (by clicking on it) from the Variables window. For the data, this window provides the file name, the number of variables that are included in the data, and the number of "observations" (e.g., survey respondents). For a given variable, the Properties window lists the variable name, its type, format, and value label. Details on each of these descriptors are discussed later in this chapter. By default, the Properties window is "locked," meaning you cannot change any of these characterstics directly from the Properties window. Clicking on the padlock icon (located in the upper left corner of the Properties window), however, unlocks the Properties window and allows you to change the aspects of the variable simply by clicking on that property (e.g., the variable name). More details on this process are provided later in this chapter.

5. Command Window: The Command window is where you will enter the operations that you want Stata to perform, when using the "syntax" interface. Syntax, or code, is another term for Stata's command language. These are the words that tell Stata what procedures to execute. Commands are entered, one at a time, in this window. After you type a command into the Command window, pressing the Enter key on your keyboard makes Stata execute the procedure that is defined by the Command. One helpful feature of the Command window is that you can scroll through previously executed commands by pressing the Page Up key. When you find the previous command you are interested in, you can either alter it or simply press Enter again to rerun the same command. The majority of this book will be devoted to explaining and describing the various commands that you will need to use to perform quantitative analyses.

There also are several icons at the top of the screen. The purpose and use of these icons are covered throughout the book. Each of these basic windows will become familiar to you as we go through this book. For now, be sure that you feel comfortable identifying the main purpose of each of the windows.

Getting Started With Data Files

When working with Stata, you will be using what is referred to as a "data file." If you are familiar with typical database programs, then you already know what a data file basically is. These files contain information (often numerical) on a set of cases, such as respondents to a survey, a sample of schools, or each of the states in the United States. Generally, data files are organized such that information regarding each case is contained in one row in the file, whereas each column represents a variable (i.e., information about that case), such as a person's gender, a school's total number of students, or a state's total square miles.

Similar to most computer files, data files come in many different types. But just like a PDF file is very similar to a word document, so too are all data files essential derivations of a similar structure. Each of these derivations is denoted by a different file extension—the letters that come after the "." in a file name. The primary file for Stata data files is .dta. Moving other types of data files into Stata (e.g., Microsoft Excel files) is covered in the Using Different Types of Data Files in Stata section of this chapter.

OPENING AND SAVING STATA DATA FILES

To open a data file that is in Stata format (i.e., one that has a .dta extension), select the File menu (in the upper left-hand corner), then choose Open. Or alternatively, you can simply click on the icon. From here you will need to search through the disk drives and folders on your computer to find your saved data file. This chapter uses the data file, available at **www.sagepub.com/longest** named `Chapter 1 Data.dta`. Once you have found your data file, double click the file. Having done this, you will notice that the Stata screen looks different from how it did initially.

The first operation you performed is now displayed in both the Results and Review windows. Again, whenever we tell Stata to "do" something, whether through the point-and-click menus or by entering a command in the Command window, it will be displayed in the Results and Review windows. Because opening a data file does not have any "results," only the command is

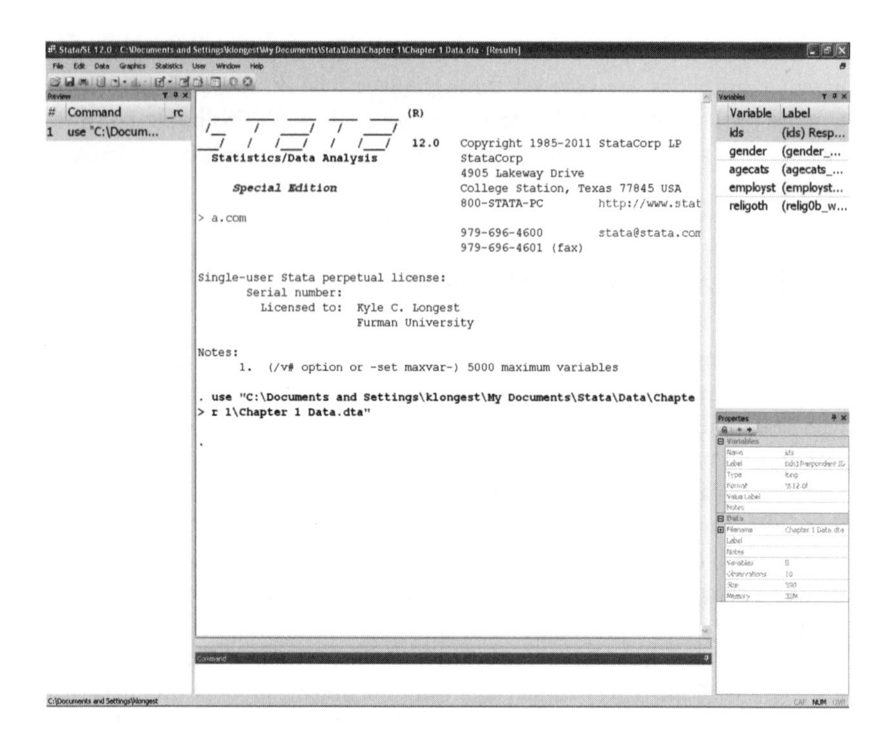

displayed in the Results window. You also can see that the data file contains five variables, listed in the Variables window. All the information provided about each variable in this window is discussed in a later section, but for now the most important aspect is the variable name. In this data set, the five variables are named `ids`, `gender`, `agecats`, `employst`, and `religoth`. These variable names should give you some indication of what type of information the variable contains. The variable gender, for example, says whether each respondent is a male or a female.

It is a good practice to always save a copy of your data files and only work with that duplicated version. When working with and analyzing data, you will often be forced to change aspects of the data files. For example, you may need to create a new variable or change something about an existing variable. But it is important to have an original version of the data, just in case something undesired occurs. Don't worry too much; most alterations you perform can be undone or recovered. Working with a duplicate copy of the data is simply an added protection.

To save a duplicate copy of the data file you have just opened, open the **File** menu and click on **Save As**. You can then enter a new file name, such as `Chapter 1 Data mycopy.dta`, and click **Save**. This is the procedure you will use whenever you want to save a new version of your data file.

A Closer Look: Stata Data Files Across Versions

As was noted in the Preface, the vast majority of Stata features and commands are similar across versions (e.g., Stata 12, 11, 10, etc.). This is true of Stata data files, by and large. All Stata data files that are created and/or saved in an older version can be read by a newer version (i.e., forward compatible). That means that if you are using Stata 12 but are working with colleagues who are using Stata 11, any files they send to you will open without a problem.

During certain upgrades, however, Stata data files cease to be "backward" compatible, meaning files saved in a newer version cannot be opened by older versions. Stata 12 happens to be one of those upgrades. If you are using Stata 12 and send a data set that you saved in Stata 12 to your colleagues who are using Stata 11, they will not be able to open it. [Note: This is not a problem if you are moving files between Stata 11 and Stata 10, as these two versions are completely compatable with each other.]

Do not despair. Stata has built in a very simple feature to overcome this problem. If you know that you want the data you are using in Stata 12 to be opened by older versions, you need to take one extra step (from the process just explained).

First, click on the File menu and then click on **Save As**. Now, use the drop-down menu in the **Save as Type** box and select **Stata 9/10 Data (*.dta)** option. The option is listed as "Stata 9/10" and not 11 because Stata Versions 8 and 9 as well as 10 and 11 are completely compatible with each other (both forward and backward), so using this option actually allows the data to be opened in any version of Stata from 8 through 12. Note that you do not need to change the file extension, it is still `.dta`. Once you have named your file, click **Save**. You will know that you have saved the data correctly when the output in the Results starts with `. saveold`, which is telling you that the file has been saved in a way that makes it readable by the previous versions. Again, note that when you save a file in this way, it can still be used in Stata 12.

DATA BROWSER AND EDITOR

If this is the first time you are working with data, it may be helpful to actually "see" the data. Even if you have experience using data, it may often be helpful to look at the data you are examining. To see the data file in Stata, you can click on the Data Browser icon, , in the middle of the top of the screen. When you do so, you will see a new window that appears as shown below:

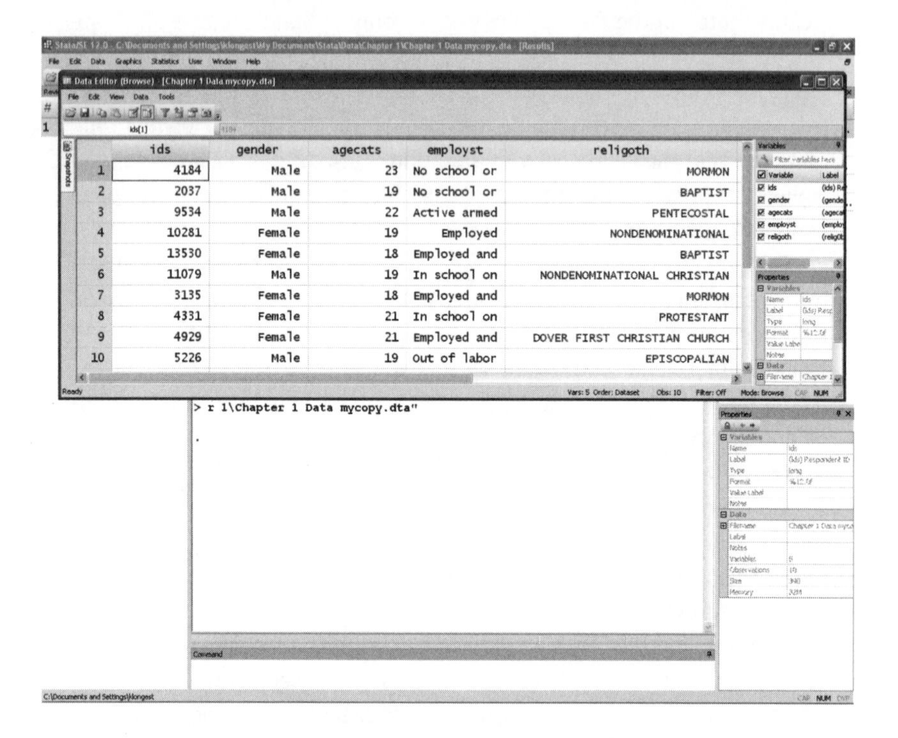

This new window, as is denoted in its upper left-hand corner, is the Data Editor (Browse) window. The "(Browse)" aspect indicates that you are only looking at the data, not actually changing them.

In this window, you see all five of the variables that were listed in the Variables window. As was mentioned earlier, each row is a different case (i.e., a National Study of Youth and Religion [NSYR] respondent), and each column is a different variable. Each cell then contains information on the given variable for that case. For example, the case in the first row is a "Male" respondent who mentioned that "Mormon" was his other religion. To close this window, click on the red "X" in the upper right-hand corner.

There may be times when you want to change the value of a particular case on an individual variable. One way to do so is by using the Data Editor window. (A more efficient way to change the values of multiple cases is covered in The 5 Essential Commands: replace (if) section of Chapter 2.) To begin, click on the Data Editor icon, , which is next to the Data Browser icon. You may notice that the Data Editor and Data Browser windows look very similar. The main difference is that in the upper left-hand corner of the window, after "Data Editor," the window now reads "(Edit)." It is important to know which window you have opened because you can change the values of the data when the Editor is open. To prevent any accidental alterations, it is generally advised only to use the Data Browser window unless you are certain you want to change a particular value.

After you have opened the Data Editor window, use the direction keys (or mouse) to highlight the cell you would like to change. For example, you may have realized that the first case's age was incorrectly entered in the data file. Instead of being 23 years old, this case should only be 22 years old. To make this change, once you have the cell in the first row listed under `agecats` highlighted, type 22 and press **Enter**. This case's value for the variable `agecats` has now changed. When you close the Data Editor window, this operation has been recorded and displayed in both the Review and Results windows.

A Closer Look: Your First Command

You may have noticed that when you changed the first case's value using the Data Editor window, the following text was displayed in the Results window:

```
. replace agecats = 22 in 1
(1 real change made)
```

Whenever you use the menus or a point-and-click method for performing an operation in Stata, it displays the command that would be entered in the Command window to perform the same operation in the Results window. In this Data Editor example, you can see that the command to change a value is `-replace-`. If you had entered this full command into the Command window and pressed **Enter**, the same change would have been made. At times, it may be helpful to perform an operation for the first time using the menus, but, as will be discussed in much more detail in Chapter 2, it is extremely beneficial to know and use the commands via the Command window for the majority of the operations you need to perform.

The rest of this book will discuss how to perform operations using the Command window. But to see the connection between the menu-based operation and the Command window, try this: Type (or copy and paste) the full command (except the first ".") that was displayed in the Results window when you closed the Data Editor window into the Command window. Now change the "22" to "23." The command should read

```
replace agecats = 23 in 1
```

Then press **Enter**. Open the Data Browser window again and notice the change to the first case's value under `agecats`.

ENTERING YOUR OWN DATA

Many data files that you will analyze will already be in Stata format or in a format that can be easily converted to Stata format (more on this topic below). Yet there may be times when you need to enter the data from a study. For example, if you distributed a survey through the mail, you will need to input the responses to each question for each case so that you can analyze them in Stata.

The first step in entering your own data after you have opened Stata is to open the Data Editor window as above. From here you can simply enter the values for each case on each variable. Entering data in this way is very similar to entering values into a Microsoft Excel file. The Data Editor, however, does not have the equation functionalities that an Excel file would.

When you begin entering values, each variable is automatically named `var1`, `var2`, and so on. Most often it is helpful to have the variable names be more descriptive of the values they contain. One way to change these generic names to something that more clearly identifies the variable is to click on the current name of the given variable you want to rename (e.g., `var1`) listed near the top of the Editor window. Doing so will bring up that variable's information in the Properties window (inside the Data Editor window). Then click on the current variable name listed in the Name blank in that Properties window. From there you can simply delete the current name and enter the desired name. Another option would be to close the Data Editor window when you have finished entering all of the data. Then you can click on the variable name (e.g., `var2`) in the Variables window, which will bring up that variable's information in the Properties window. To change the name in this Properties window, you will need to click on the padlock icon in the Properties window. Then you click on the current variable name listed in the Name blank and simply type the new name in the blank.

Once you have finished entering all of your data, close the Data Editor and follow the steps described above to save a copy of your data file in Stata format.

USING DIFFERENT TYPES OF DATA FILES IN STATA

Some data files may not be available in Stata format. Therefore, a few steps are needed to work with these files in Stata. It would be virtually impossible to cover every possible data file type and how each can be transferred to be usable in Stata. Instead, the most common type will be covered. Also note that there are other computer software programs that are specifically designed to convert data files into various formats (e.g., Stat/Transfer). If you have access to such a program, it is probably the most effective and efficient way to transfer files into a Stata format. Some statistical software packages also offer the option of saving a data file in a different format, which often includes the Stata, .dta extension.

One of the most frequently encountered data file type that is not Stata-ready is a Microsoft Excel file. Usually these files are denoted with the .xls extension, but other extensions (e.g., .csv) that are generated or readable by Microsoft Excel can all be treated in a similar fashion.

This process requires that you have access to and some familiarity with Microsoft Excel. To start, open the data file in Microsoft Excel. Then highlight the entire worksheet that contains the data and copy it (either by right clicking and choosing Copy or using the copy function (Ctrl+C)). Next, in Stata open the Data Editor window, highlight the upper left data cell, right click and choose Paste, or use the paste function (Ctrl+V). Once you pasted in the data, you should be presented with a window that asks whether you want to Treat First Row as Data or Treat First Row as Variable Names. The option that you choose will depend on whether your Excel file contains variable names in the first row or whether it contains only data. The two formats are shown below.

First Row as Variable Names

	A	B	C	D	E
1	ids	gender	agecats	employst	religoth
2	4184	Male	23	No school or	MORMON
3	2037	Male	19	No school or	BAPTIST
4	9534	Male	22	Active armed	PENTECOSTAL
5	10281	Female	19	Employed	NONDENOMINATIONAL
6	13530	Female	18	Employed and	BAPTIST
7	11079	Male	19	In school on	NONDENOMINATIONAL CHRISTIAN
8	3135	Female	18	Employed and	MORMON
9	4331	Female	21	In school on	PROTESTANT
10	4929	Female	21	Employed and	DOVER FIRST CHRISTIAN CHURCH
11	5226	Male	19	Out of labor	EPISCOPALIAN
12					
13					
14					
15					
16					
17					
18					
19					

First Row as Data

	A	B	C	D	E	F
1	4184	Male	23	No school or	MORMON	
2	2037	Male	19	No school or	BAPTIST	
3	9534	Male	22	Active armed	PENTECOSTAL	
4	10281	Female	19	Employed	NONDENOMINATIONAL	
5	13530	Female	18	Employed and	BAPTIST	
6	11079	Male	19	In school on	NONDENOMINATIONAL CHRISTIAN	
7	3135	Female	18	Employed and	MORMON	
8	4331	Female	21	In school on	PROTESTANT	
9	4929	Female	21	Employed and	DOVER FIRST CHRISTIAN CHURCH	
10	5226	Male	19	Out of labor	EPISCOPALIAN	

After you have selected the option that fits with the type of data file you have, close the Data Editor, and follow the previously described steps to save the data from within Stata as a Stata data file. Once you have saved your data as a Stata data file, you can simply open and use this version of your data.[3]

Stata 12 (but not the previous versions) offers another method for bringing data from an Excel file into Stata that may be even slightly quicker. After opening Stata, click on the **File** menu, followed by **Import**. Select the **Excel spreadsheet (*.xls *.xlsx)** option,[4] and the following window will appear:

---

[3]This "copy and paste" method is the easiest way to transfer data from Microsoft Excel into a Stata format, especially for novice users. But there are some disadvantages to this strategy. More practiced users should transform Excel worksheets into .csv files and then implement the `-insheet-` command. The specifics of this command are beyond the scope of this introductory text, but the Stata Help Files section of Chapter 8 provides information on how Stata's Help files can be used to learn how to use this command.

[4]If you are using Stata 12, you will also notice that you could select several different data file formats from this window. The general procedure for each of these formats is very similar to the procedures discussed for Excel files but may contain particular steps for specific files.

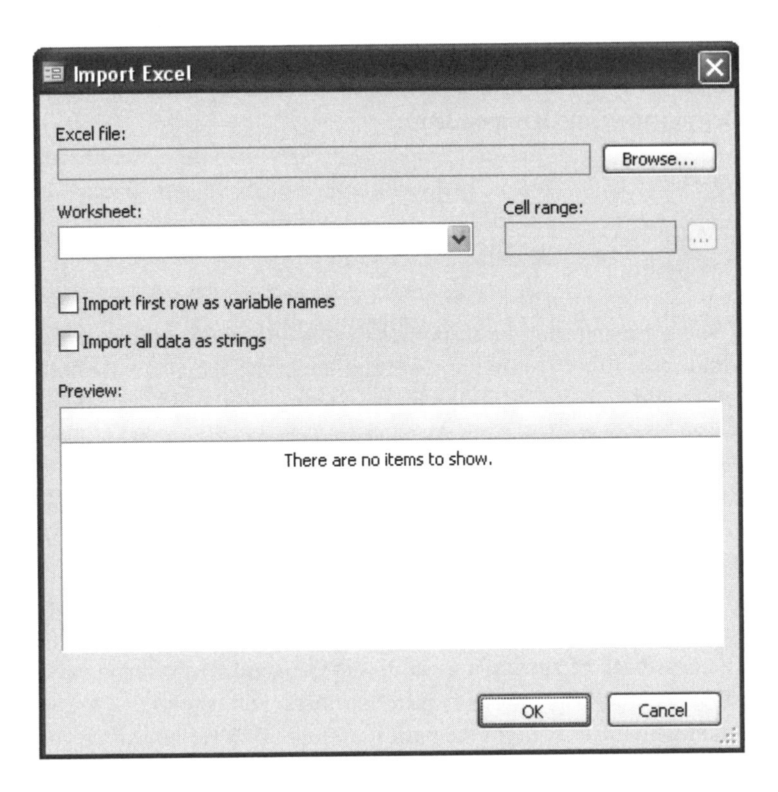

Click on the **Browse** button to find the Excel data that you would like to turn into a Stata data set. Once you have selected the Excel file, you can pick a particular worksheet from that file or even a particular set of cells by using the corresponding boxes. Notice that you still need to decide and tell Stata whether the first row in the Excel file contains variable names or actual data. If the first row contains variable names, click the radio button next to **Import first row as variable names** (when you do this, notice that the data shown in the preview window will change). Then click **OK**. As described above, you can follow the previously described steps to save the data from within Stata as a Stata data file. Once you have saved your data as a Stata data file, you can simply open and use this version of your data.

TYPES OF VARIABLES IN DATA FILES

At this point, you should feel comfortable with the basic structure of data files. Each row holds the information for one case and each column is a different variable. With this knowledge, you are almost ready to start analyzing your data. There is, however, one distinction in the types of variables included in data that is important to understand.

To help illustrate this difference, consider the NSYR variable, in the `Chapter 1 Data.dta` file, `gender`. This variable came from the following question asked of all respondents:

Are you

a. Male?
b. Female?

If you were entering the responses to this question into a Stata data set, you could record them in one of two ways. First, the actual answer "Male" or "Female" could be recorded for each case. Second, you could use a number to represent each answer. For example, you could choose to enter 0 for all respondents reporting "Male" and 1 for all respondents reporting "Female."

If you record the responses in the first way, it would be what Stata refers to as a *string* variable. A string variable is a variable in which the contents are actual words. String variables can be very useful for many purposes. For example, you can enter verbatim answers to questions directly into Stata, as was done for the variable `religoth` in the `Chapter 1 Data.dta`.

The drawback of storing a variable such as gender as a string variable is that some statistical operations require numbers. For example, if you wanted to calculate the mean (i.e., mathematical average) of a variable, each category must be assigned a numeric value. For this reason, it is generally advisable, when possible, to use the second method and enter variables as *numeric* variables. These are variables that have actual numbers attached to each response.

Fortunately, many of the Stata commands that will be discussed in this book operate similarly with numeric or string variables. The commands that work only with numeric variables are those that perform statistical operations that require numbers to calculate, for example, the mean or a linear regression. Because numeric variables, typically, are more applicable to the vast majority of data analyses, the commands discussed in this book focus on their use with numeric variables (keeping in mind that many operate identically for string variables). The primary commands that are used (and are different) for string variables, including methods for changing a string variable to a numeric variable, are addressed in the Data Management: Using String Variables section in Chapter 3.

As has been discussed, often you may be using data that you did not enter, so you may not have a choice or even be certain about the way in which variables were entered. There are several ways to determine whether a variable is a numeric or string variable. The most straightforward way is to open the Data Browser window. In versions Stata 10 or later, string variables are shown in a red font, whereas numeric variables are shown in either black or blue font. In the `Chapter 1 Data.dta` file, you will see that only the variable `religoth` is a string variable.

Another option in Stata 12 to see which variables are string variables is to click on a particular variable in the Variable window. In the Properties window, you will see an entry for **Type**. When the variable type starts with the letters "str," the variable is stored as a string variable.

A Closer Look: Variable Types

You may have noticed that more information about the variable type is listed in the Properties window. For example, `gender` is shown to be a byte variable, `ids` is a long variable, and `religoth` is a str31 variable.

These distinctions further demarcate variables within the general categories of numeric and string. They also are related to how much file space is allotted to storing the variable.

All string variables have the "str" prefix, and the number indicates the maximum characters that can be used for that string variable. So the maximum length a denomination could be in the variable `religoth` is 31 characters. As you will see, this constraint can be altered, but it is advisable to use only the minimum number of characters that are needed. Otherwise you are using memory to store empty spaces.

Similarly, the various subtypes of numeric variables indicate the number of digits that each variable can hold. In order of smallest to largest, the numeric variable types are byte, int, long, float, and double.

Generally, Stata will store variables in the most efficient and effective way when you create them. Moreover, most users of Stata will conduct countless analyses without ever having to worry or manipulate these specific distinctions.

When you have the Data Browser open, you probably notice, however, that the variables `gender` and `employst` look different from the variables `ids` and `agecats`. This difference is due to the fact that `gender` and `employst` have what are called *value labels* attached to them. Value labels will be covered in much more detail later, but they are labels that can be applied to the numeric codes used to represent responses. Remember that you could decide to use the number 1 to represent the answer "Female." This choice may be difficult to remember (i.e., whether 1 was Male or whether 1 was Female), therefore you can use value labels as a shortcut to help remember this coding strategy. The variables `ids` and `agecats` were numerical responses so they do not have any value labels that could be attached to them. You can see the actual numerical codes for each variable using the Data Browser window by clicking on the

Tools menu, selecting **Value Labels**, and clicking **Hide All Value Labels.** When you do so you will see the cases that were "Male" now display "0" and the cases that were "Female" now display "1."Or you can highlight (either using the direction keys or the mouse) a particular cell (e.g., "Male"). When you do so, the actual value is listed in a pane just underneath the icons.

Exercises

1. Open the "`Chapter 1 Exercise Data.dta`" data file.
2. Save a copy of the open data named "`Chapter 1 Ex mycopy.dta`."
3. Using the Data Browser, determine how many cases and variables are in the data set.
4. Which of the variables is a string variable?
5. Use the Data Editor to change the `agefstdt` value of the last case from 14 to 13.
6. Use the Data Editor to input an additional case with the following characteristics: ids value of 1004, is Male, completed the 12th grade, went on his first date at 16, lives in the Pacific census region, and does not live with his parents.

2

The Essentials

Now that you are familiar with the basic components of Stata and data files, it is time to begin performing statistical analyses. The thought of conducting statistical operations on top of learning a new computer program can be a doubly daunting task—often bringing with it a considerable amount of anxiety. This chapter is explicitly designed to help alleviate this common and natural emotional reaction to learning Stata to conduct statistical operations. This chapter has three primary goals. First, it presents a conceptual approach to learning Stata commands that has been shown to not only help learn the necessary operations but also assuage the fears of "memorizing" seemingly endless commands. Second, the basic structure or format of Stata commands is covered. Regardless of whether the actual operation that a command performs is straightforward or complex, all Stata commands follow a very similar structure. Knowing this underlying format will help you process each newly presented operation more easily. Finally, this chapter discusses the 5 essential commands of Stata. These 5 commands form the foundation of statistical and data management operations for the vast majority of research projects. Therefore, once you have completed this chapter, you will have mastered a significant portion of using Stata to accomplish your research. Doing so will hopefully minimize anxiety and increase confidence when approaching the more nuanced topics covered in the subsequent chapters.

Intuition and Stata Commands

Perhaps one of the more intimidating aspects of Stata is that it operates primarily, and most effectively, using a syntax, command-driven interface. As most readers have become accustomed to a Windows, point-and-click interface, this more "DOSesque" system may be unfamiliar and unusual. Furthermore, many users may be disheartened by the thought of trying to memorize numerous, odd-sounding commands.

These very valid concerns are why this book uses a new approach for teaching Stata commands. This method is founded on the idea that instead of viewing Stata as some black box that only spits out the right results when told exactly what to do, it is more beneficial to see Stata as an extremely smart colleague who you are asking to produce some calculations very quickly. The latter perspective will help you remember that although Stata is a statistical, computer program, it is designed by people. When these people thought about what to call particular commands, they did their best to give them names that made sense.

Taking the latter approach helps facilitate a more intuitive approach to Stata. Rather than considering the numerous command names that need to be "memorized," it is more effective to think as if "what would I call a command that would tell a computer to do a cross-tabulation?" Or alternatively, you can think "if my colleague and I had been working together for a long time, how might I tell him or her that I needed a cross-tabulation in a shorthand way." Generally, thinking in this manner leads you to the correct answer `-tabulation-` or `-tab-` for short. This intuitive approach should help you learn and retain Stata commands more easily and effectively. It should also help minimize worries about the prospect of using Stata.

Finally, there are times when this type of thinking may not lead to exactly the right command. For example, if you thought "what would I tell my colleague if I wanted him or her to erase a variable from the data set," you may think "erase" or "delete." The actual command for this operation is `-drop-`. But approaching the new command in this way should lead you more quickly to the correct command and will help make the actual command make more sense and be more easily remembered.

Thus, as we embark on learning all the wonderful things Stata can do, keep this intuitive approach in mind. Remember you are simply working with a really smart colleague. Sometimes communication may become strained, but with a bit of dialogue and understanding, you will be able to conduct very effective analyses.

A Closer Look: Commands versus Point-and-Click

Often new Stata users are apprehensive about using Stata because of its command-driven interface, rather than a Windows, point-and-click–based system. Sometimes this concern may tempt users to disregard learning Stata commands and instead rely solely on its Menus and point-and-click operation. Although this path may seem appealing, there are several reasons to fight the urge.

First, using the point-and-click method is not any easier, in terms of the amount of information you need to know. That is, even when using a Windows-based program, you still need to learn which menus to open, which button to use for a particular operation, and the correct options to choose. This method may seem easier than learning the commands, but it is not due to a difference in quantity of information to be attained. The distinction in the two methods rests mainly in the familiarity with using menus and Windows to perform operations. But at one time this method was probably intimidating as well. Just as many people have come to feel very comfortable using Windows-based computer programs, with a little practice, the Stata syntax, command-based interface will seem just as straightforward.

Second, and perhaps even more important, there are real advantages to knowing the command-based aspect of Stata. For the majority of operations, the command-based interface is much quicker than the menus. What can take several point-and-clicks to, process through the necessary layers of options, usually can be typed in a few short words. Furthermore, although similar operations can be performed using either method, the command-based format makes it much easier to save and replicate your data manipulation and analyses. Often, you need to make adjustments to previously conducted procedures and run them again. As will be shown in the What Is a Do File? section of Chapter 3, using the commands along with "do files" makes this process much more straightforward. Additionally, if you continue to use Stata, many of the more advanced abilities of the program rely on the command format.

The Structure of Stata Commands

This section provides an overview of the components of Stata commands. Much more detail and specific examples are covered throughout the chapter, which will help clarify each aspect. Every command that is performed in Stata has the same basic structure, which can be written in generic terms as follows:

```
command varname(s) [if varname==value] [, options]
```

COMMAND

Any statistical or data operation you want to perform in Stata has a name. For example, if you would like to delete an entire variable from the data, the command would be `-drop-`. These commands are generally the first item that is typed in the Command window (or "do" file, covered in Chapter 3).

Most commands have two forms: a full command name and an abbreviated command name. The abbreviated command name contains the minimum number of characters required to uniquely specify that command. If a command has an abbreviation, you can type as many of the characters as you would like, as long as it contains the minimum abbreviation. For example, the full command to perform a linear regression is `-regress-`, but the abbreviated command name is `-reg-`. Therefore, you could type `-regress-`, `-reg-`, `-regr-`, `-regre-`, or `-regres-`, and the same operation would be performed. This book always introduces a command using the full command name, but often an abbreviated command name is used for the sake of simplicity after this first use.

VARIABLES

After the command, you must specify the variable or variables on which you want to perform that operation. For example, if you wanted to delete a variable named `gender`, you would type `drop gender` into the Command window. Particular commands, as will be discussed in more detail, either accommodate multiple variables or even require multiple variables to be specified. If, for instance, you wanted to create a cross-tabulation, you would specify two variables after the appropriate command.

IF STATEMENTS

There may be times when you want to perform an operation only on certain types of cases. As an example, you may want to produce a cross-tabulation table that includes only the males in your data set. To do so, you would type an `if` statement after you have entered the command and variables. Generally, these `if` statements take the form of a particular variable or variables equaling some value.

The `if` statements are completely optional; meaning you do not have to enter them when performing a command, which is why they are shown in brackets above. You need to type an `if` statement only when you wish to perform the operation on a selected set of cases in the data.

OPTIONS

Most Stata commands include options that can be invoked with them. As the name suggests, option statements are optional. Options perform some extension or modification of the basic command, such as requesting additional statistical measures or a different formatting of the output. When a Stata command does not produce exactly what you would like by default, you often can obtain what you are looking for through the use of options. When each command is covered throughout the book, the most helpful options will be detailed as well. Furthermore, the Stata Help Files section of Chapter 8 shows how to learn all the possible options for each command.

EXECUTING A COMMAND USING THE COMMAND WINDOW

Once you have determined which command you need to use, which variables you want to perform it on, and whether you would like to use an `if` statement or options, you are ready to execute the command.

First, be sure that you have the Command window selected by clicking the mouse when the cursor is anywhere in the Command window. Next, you will type the command, variable name(s), and any desired `if` statements or options. Instead of actually typing a variable name, you can also place your cursor over a particular variable in the Variables window, and when you click on the little arrow that appears, its name will appear in the Command window. If you are using Stata 11 (or earlier), you can simply click on the variable name in the Variables window, and that variable's name will appear in the Command window. Once you have finished entering all the information, press **Enter**.

Pressing **Enter** tells Stata to perform the operation and causes output to be displayed in the Results window. Note that some commands may wrap onto more than one line in the Command window. This scenerio is completely acceptable. Stata treats everything typed before you press **Enter** as a single command. Thus, for each command you wish to perform, you need to type all the required information and press **Enter** (i.e., you cannot type several commands in succession in the Command window). A method for performing multiple commands at once is covered in the next chapter.

The 5 Essential Commands

The following section provides a closer look at the foundation commands of Stata. These 5 commands accomplish a significant portion of the analyses and data management that is needed for many research projects. This section should be seen, however, as an introduction to these commands. It explains the basics of each command, which for many users may be all that is needed. More of the specifics and nuances for each command are covered in the chapter devoted to that particular statistical operation. Therefore, the goal of this section is threefold. First, it provides essential commands that perform some of the most frequently used operations. Second, it gives you a framework on which to place all of the more advanced topics to come in later chapters. Third, it should give you confidence to tackle those more advanced topics. When you grasp these core concepts, you are in a great position to become an effective Stata user.

All the examples that follow use the `Chapter 2 Data.dta`, available at **www.sagepub.com/longest**, This data set contains 7 variables for 25 cases from the National Study of Youth and Religion (NSYR) data (see the Preface for more information on how these data were collected). This subsample of the full data was selected so that it would be possible for you to double-check the following analyses by producing it by hand if it is helpful. As mentioned in Chapter 1, it is a good idea to save a copy of the data file you are working with so that you always have a backup of the original data.

tabulate

The first two essential commands, `-tabulate-` and `-summary-`, both produce basic descriptive information about variables, which is why they generally are the first analytic operation performed for the vast majority of research studies. Again this section will provide more of an overview on how to use these commands, whereas Chapter 4 presents much greater detail on the specifics of using these commands, as well as more detail on extensions to each.

One of the first analytic processes taught in statistics courses is how to construct a frequency distribution table. Notice, if you were asking a really smart colleague to produce a distribution that tabulates the values of each of the cases, you might tell him or her to "tabulate" the data. The abbreviated command name for `-tabulate-` is technically `-ta-`, but it will probably be easier to remember `-tab-` as a shortened version.

To see what the `-tab-` command does, select the Command window (i.e., click the mouse while the cursor is over the Command window), and type `tab employst` (or alternatively, type `tab` and place your cursor over `employst` in the Variable window and click the small arrow that appears). The `employst` variable comes from a question asking about the respondents' current employment status. When you have typed the command, the screen should look similar to the screen shot presented below:

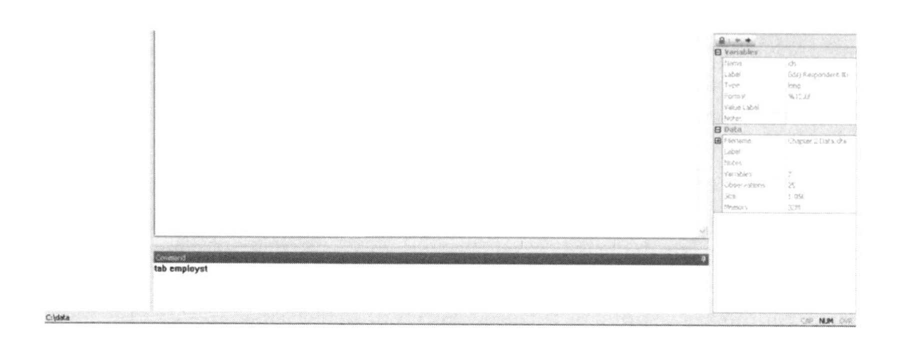

Now simply press Enter, and your screen should look like this:

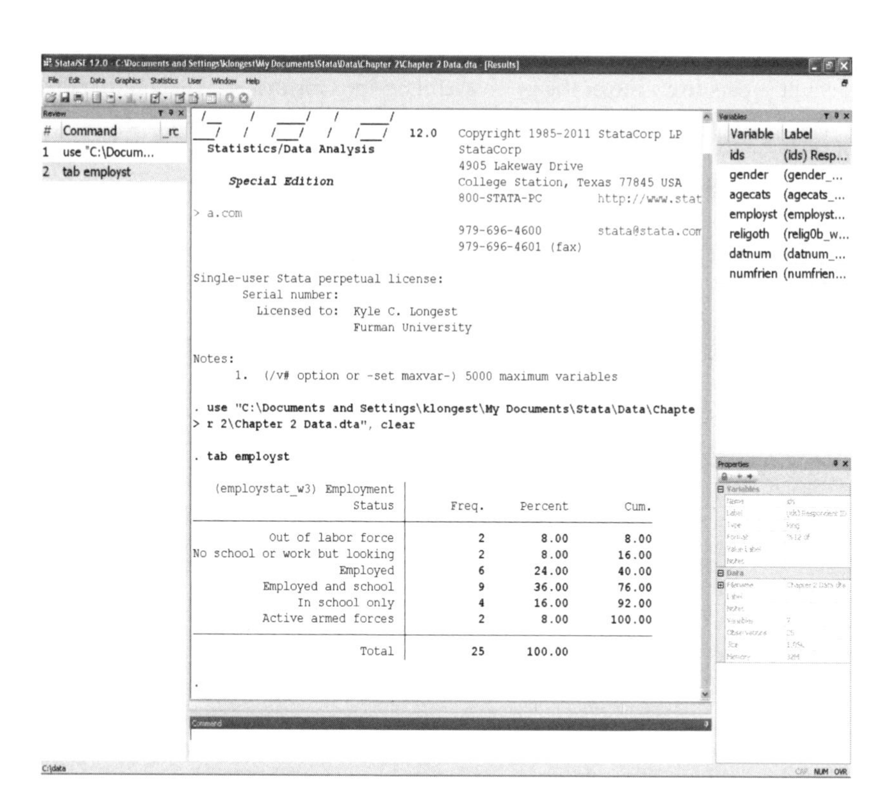

Before addressing the unique aspects of the information provided by the -tab- command, examine some of the general output that is produced by all Stata commands. First, you see what the command "did" displayed in the Results window. In this case, this is a distribution table showing the frequency, percentage, and cumulative percentage for each category of the variable employst. Next, just above this output, Stata presents the exact command that was executed to produce the output. This same information is also stored in the Review window. These three components are produced for every command you enter in the Command window.

Turning to the output of the -tab- command, as noted, it produces a table showing the frequency, percentage, and cumulative distribution of the given variable. For the employst variable, six cases are Employed, which is 24% of the sample. Also notice that it displays the total number of cases that fall into at least one of these categories of the variable. In this case, there are 25 cases that were coded into one of the presented categories.[1] The top left corner of the table lists what is called the "variable label." These labels usually provide a brief description of the variable, such as "Employment Status." Working with these labels is covered in more detail in Chapter 3.

As with all commands, the -tab- command contains several options that can be invoked to perform different or additional operations beyond this default procedure. One of the more useful options for the -tab- command is -sort-. The -sort- option tells Stata, as it sounds, to rearrange (i.e., sort) the table so that it lists the categories in descending order of frequency. Remember, options are always typed after entering a comma, meaning you would type tab employst, sort in the Command window and press **Enter**. When you do this, the following output is presented:

```
(employstat_w3) Employment |
                    Status |      Freq.     Percent        Cum.
---------------------------+-----------------------------------
       Employed and school |          9       36.00       36.00
                  Employed |          6       24.00       60.00
            In school only |          4       16.00       76.00
         Out of labor force |          2        8.00       84.00
No school or work but looking |       2        8.00       92.00
        Active armed forces |          2        8.00      100.00
---------------------------+-----------------------------------
                     Total |         25      100.00
```

[1]For these introductory examples, none of the variables have any missing data, meaning all of the cases have valid answers for all of the variables. Clearly this situation may not always be the case with real data. Handling such missing data will be covered in more detail in the later chapters that more thoroughly discuss using the commands to complete statistical analyses. Specifically, see the Data Management-Missing Data section of Chapter 3.

As you can see, the same basic information (frequency, percentage, and cumulative percentage) is displayed, but now the categories are ordered so that you can easily see which one contains the most respondents and which contains the least. For employment status, the most common category is Employed and school, whereas being Out of labor force, No school or work but looking (for a job), and Active armed forces are all tied for the least common responses.

There are several other options you can use with `-tab-` and most of them are covered in the Frequency Distributions section of Chapter 4. But for now, it is only necessary to understand the basic form of how options are invoked, as it is similar for all other commands. Also, it should be noted that as with commands, options have full names and abbreviated names. The full name will always be presented when it is first introduced, with the most common abbreviation being used in all the following instances.

In addition to producing a distribution of one variable, the `-tab-` command can generate a cross-tabulation between two variables. For instance, you may be interested to know if there is a difference in the respondents' employment status by gender. To make this comparison, you would want to see the distribution of employment status for males and the distribution of employment status for females. One method for displaying this information is to invoke the `-tab-` command and list both variables instead of just one. Type `tab employst gender` in the Command window and press **Enter**. Doing so produces the following results:

```
                           | (gender_w3)Respondent
          (employstat_w3)  |        gender
         Employment Status |      Male      Female |     Total
---------------------------+----------------------+----------
         Out of labor force |         0           2 |         2
No school or work but looking |      0           2 |         2
                  Employed |         2           4 |         6
       Employed and school |         3           6 |         9
            In school only |         3           1 |         4
       Active armed forces |         1           1 |         2
---------------------------+----------------------+----------
                     Total |         9          16 |        25
```

From this table, you can see that there are more females (16) than males (9) in this sample. Additionally, females are more prevalent in all the categories, except in the In school only and Active armed forces categories. But these differences in frequencies could stem from the greater number of females overall. Therefore, to know if there is a true relationship, you would want to know the percentage of females in each category compared with the percentage of males in each category.

To produce the necessary figures, you can again think intuitively. You need to ask Stata to produce a set of percentages based either on the rows or on the columns. Because you believe that gender is the causal variable (i.e., the independent variable), you would want the percentages in the columns. That is, you want to be able to compare the proportion of *all females* who are employed with the proportion of *all males* who are employed. Therefore, the percentages need to be calculated within the columns. To have your smart colleague make this calculation, you might consider telling him or her, for shorthand, "columns." Following this logic, the option to present these percentages is `-column-`. [If you wanted the percentages in the row, as you might have guessed, the option would be `-row-`.] Type `tab employst gender, col` in the Command window and press **Enter**.

```
+-------------------+
| Key               |
|-------------------|
|     frequency     |
| column percentage |
+-------------------+

                        | (gender_w3)Respondent
       (employstat_w3)  |         gender
    Employment Status   |      Male       Female |     Total
------------------------+------------------------+-----------
   Out of labor force   |         0            2 |         2
                        |      0.00        12.50 |      8.00
------------------------+------------------------+-----------
No school or work but looking |   0            2 |         2
                        |      0.00        12.50 |      8.00
------------------------+------------------------+-----------
             Employed   |         2            4 |         6
                        |     22.22        25.00 |     24.00
------------------------+------------------------+-----------
  Employed and school   |         3            6 |         9
                        |     33.33        37.50 |     36.00
------------------------+------------------------+-----------
       In school only   |         3            1 |         4
                        |     33.33         6.25 |     16.00
------------------------+------------------------+-----------
  Active armed forces   |         1            1 |         2
                        |     11.11         6.25 |      8.00
------------------------+------------------------+-----------
                Total   |         9           16 |        25
                        |    100.00       100.00 |    100.00
```

A key is displayed at the top of the table to indicate what each of the numbers in each cell represent, in this case the frequency and the column percentage. Using the percentages you can more accurately compare the relationship of gender on employment status. When only examining the frequencies, it appeared that females were more likely to be employed because 4 females reported being employed compared with only 2 males. But using the percentages, it now appears as if there is no difference by gender in terms of being employed because 25% of females are employed and just over 22% of males are employed, a difference of less than 3%.

A Closer Look: Command Shortcuts

You might have noticed that the three previous commands were very similar, with the latter two simply adding a new option. You may wonder if there is a shortcut to reproduce these commands and add the new option without having to type out the entire command again. Fortunately, there are actually two methods to save you from this repetitive typing. To practice both methods, consider if you wanted to see the percentages based on the rows in addition to the columns.

The first method uses the Review window. To start, find the command that is closest to what you would now like to perform. In this case, you would select the command that first produced the column percentages (i.e., `tab employst gender, col`). Move your cursor onto this command and click on it. The command appears in the Command window. Now you can select the Command window, type -`row`- to the end of the command, so that it now would read `tab employst gender, col row`, and press **Enter**.

For the second method, be sure that you have the Command window selected (i.e., your cursor should be flashing in the Command window). Then press the **Page Up** key. The command you just ran is now displayed in the Command window. If you continue pressing the **Page Up** key, you will cycle through all the commands you have invoked during the current session. You can find the one you are looking for, type -`row`- at the end and press **Enter**.

While both of these methods are efficient shortcuts for rerunning commands during one session of Stata, the What Is a Do File? section of Chapter 3 explains an even more effective method for running similar commands over multiple sessions of Stata.

summary

Often you need to analyze variables that contain numerous categories, such as income or number of people in a county (i.e., interval-ratio variables). For such variables, it would be possible to produce a distribution table, but it might not be very useful because each category would most likely contain only a few cases. In the current data set, the variable `-datnum-` comes from a question on the NSYR that asked how many romantic relationships the respondent has been involved in during his or her lifetime. The question used an open-ended response format, so that respondents could report any number between 0 and 100. With only 25 respondents in the current subsample, it is likely that each case has its own unique value, making a distribution table not very helpful.

In such a case, it would be more helpful to see figures that describe the basic pattern of values without actually seeing each case's specific response. Thinking intuitively, you would be asking your smart colleague to "summarize" the variable with a few numbers. These summary statistics are often referred to as measures of central tendency and variability. The command to produce these statistics is `-summary-`, or `-sum-`. To produce these figures, type `sum datnum` in the Command window and press **Enter**. Doing so presents the following results:

```
Variable |       Obs        Mean    Std. Dev.       Min        Max
---------+--------------------------------------------------------
  datnum |        25        7.56    6.083311          1         20
```

The default `-sum-` command (i.e., with no options included) produces several descriptive statistics. Moving from left to right: the number of observations with data on the given variable, its arithmetic mean, the standard deviation, and the smallest (minimum) and largest (maximum) value any case reports for that variable are presented. The figures are all based on observations with information on that variable in the current data set. For example, respondents in the NSYR could have possibly reported having dated up to 100 people, but no one in the current subsample claimed to have dated more than 20 people.

One helpful feature of the `-sum-` command is that it allows for more than one variable to be entered in a single command line. For example, try entering `sum datnum agecats` in the Command window and press **Enter**. The following output will be displayed:

```
Variable |       Obs        Mean    Std. Dev.       Min        Max
---------+--------------------------------------------------------
  datnum |        25        7.56    6.083311          1         20
 agecats |        25       20.04    1.619671         18         23
```

The same information is displayed, but now the descriptive statistics for both number of people dated and age are shown. The -sum- command accepts as many variables as you would like to enter.

There may be times when you would like to know additional descriptive statistics besides the ones displayed by the -sum- default, such as the percentile values or the kurtosis. To have Stata produce these statistics, you can invoke the -detail- option with the -sum- command. In the Command window, type sum datnum, detail and press **Enter**. Now the results appear as shown below:

```
            (datnum_w3) [IF HAS BEEN IN A ROMANTIC
           RELATIONSHIP OR HAS BEEN MARRIED J:4. How
-------------------------------------------------------------
      Percentiles      Smallest
 1%          1            1
 5%          2            2
10%          2            2      Obs              25
25%          3            2      Sum of Wgt.      25

50%          5                   Mean            7.56
                    Largest      Std. Dev.       6.083311
75%         10           15
90%         20           20      Variance        37.00667
95%         20           20      Skewness        .9859665
99%         20           20      Kurtosis        2.722225
```

Now, in addition to the observations, mean, minimum, and maximum, the -sum- command has produced multiple percentile values, the variance, skewness, and kurtosis score. Many users may be interested in the median value of particular variables. Although Stata does not indicate a value as the "median," the figure listed as the 50% value is equivalent to the median value. Just as with the default -sum- command, you can enter multiple variables in one command line, and a detailed descriptive statistic table for each one will be displayed.

generate

Frequently when you are working with data, you need to generate new variables that do not initially exist in the data set. For example, you might realize that using the number of people a respondent has dated is a problematic figure because the respondents are all of different ages. Therefore, people who are older would presumably have had a longer time to have dated more people. To adjust for this possibility, you decide it is necessary to create a variable that represents the number of people the respondent has dated per year, since he or she turned 16. To generate this variable, you would need to divide the total number of people the respondent reports to have dated by the number of years that have passed since he or she has turned 16. Unfortunately, there is not a variable in the current data set that holds the latter information. Therefore, you need to create two new variables in order to achieve your goal. First, you have to generate a variable that represents the number of years over 16 a person is and then divide the total number of people dated by that new variable.

Intuitively, if you were going to tell your smart colleague to make a new variable, what single word command would you use? Perhaps you might consider "create," which makes sense, but another similar option would be "generate." If you initially thought "create," remember that although your intuition may not have led you to the exact right command name, thinking in this way should now help you remember the correct one: `-generate-`. As opposed to the previous commands you have learned, `-tab-` and `-sum-`, the `-generate-` command (abbreviated `-gen-`) is not a data analysis command. That is, it does not produce results of statistics or figures. Rather `-generate-` is a data management command that is used to create new variables that you can then analyze. But remember all Stata commands follow a similar structure.

So the first word you will type in the Command window is the command name `-gen-`. Next you need to type what the newly generated variable name will be. Although you can name this new variable just about anything you would like, there are a few general tips when deciding on a new variable name. As mentioned in the Preface, Stata is case-sensitive. This feature is why it is an effective strategy to use variable names that contain only lowercase letters. It is possible to name a variable `gender`, `Gender`, or even `GENDER`. But there are times at which you will need to type this variable name. It is generally quicker, easier and creates less chance for mistakes to only use lowercase letters in variable names. Next, try to be as succinct as possible. You might have to type this variable name several times, so the shorter the better. But you must balance this succinctness with a concern for clarity. It might be tempting to call your new variable something like `newvar1`. Although you might remember what that variable represents right now, after creating several new variables it might get confusing. This scenario is why it is best to give your new variable a name that will help you remember exactly what it represents. The variable that you need to create first in the given example is going to indicate the number of years past 16 the respondent is. Therefore, you might call this new variable `agep16`. This variable name tells you that the variable is the person's age "post" 16. Again, you should name the variable whatever will help you remember what it contains using the fewest characters possible.

Once you have decided on the new variable's name, type it into the Command window. As of now, the command should read `gen agep16`. Next, you have to tell Stata (or your smart colleague) what exactly should be in this new variable. Or, in other words, you need to indicate what this new variable will *equal.* Thus, after the new variable's name you need to type an equal sign, making the command `gen agep16=`. Do not worry about whether there are spaces before or after the equal sign, Stata does not care about how many or few spaces there are in a command line. From here, you can probably determine that the remainder of the command line should indicate exactly what the new variable will contain. This portion of the command is similar to a formula or equation and can contain just about any operation you can think of.

Most often, what follows the equal sign in a generate command is one of or some combination of three elements: numbers, mathematical functions, and existing variables. It is important to remember that whatever you type after the equal sign is applied to every case in data set (unless you invoke an `-if-` statement, which is described next). That is, in a way, Stata goes through each case, one-by-one, and executes the formula you type after the equal sign. For example, try typing `gen examp=200` in the Command window and press **Enter**. Notice that in the Variables window there is a new variable listed with the name `examp`. Now produce a distribution of this new variable (i.e., type `tab examp` in the Command window and press **Enter**). The results should be as follows:

```
      examp |      Freq.     Percent        Cum.
------------+-----------------------------------
        200 |         25      100.00      100.00
------------+-----------------------------------
      Total |         25      100.00
```

This table shows that all 25 cases have been assigned a value of 200 for this new variable.

For the purposes of the substantive example, you need a variable that represents the number of years after 16 a respondent is. To create such a variable, you would need to subtract 16 from the person's current age. Using numbers alone, you might consider typing: `gen examp2=21-16`, to generate a variable that will represent the number of years past 16 for all 21-year-olds. If you execute this command and again produce a distribution of it, the results should be displayed as follows:

```
     examp2 |      Freq.     Percent        Cum.
------------+-----------------------------------
          5 |         25      100.00      100.00
------------+-----------------------------------
      Total |         25      100.00
```

This new variable contains the correct information for what was typed. A person who is 21 is 5 years past 16-year-olds. But you can see that it has assigned 5 to every case. If you look back at the summary table you produced earlier for the `agecats` variable (to do so simply use the scroll bar on the right side of the Results window or the **Page Up** key after selecting the Results window), you can see that not everyone in the sample is 21 years old. At least one of the respondents is 18 years and another 23 years, meaning that these cases would be 2 and 7 years past 16, respectively.

To produce a variable that contains the correct information for each case, you need to tell Stata to generate a new variable that is the case's actual age minus 16. Instead of putting "21," as you did for the last example, you need Stata to use each case's actual age and then subtract 16. Which variable contains this information that you already have in Stata? The variable `agecats`. Thus, the correct command to produce a variable that represents each person's number of years past 16 would be `generate agep16=agecats-16`. Notice, you simply replace "21" with the variable name that contains each case's actual age. Stata now knows to go through each case, one-by-one, and take that case's value in the `agecats` variable, subtract 16 and then enter this into the new variable. Type this command into the Command window and press **Enter**. Then, display a distribution table of this new variable (i.e., `tab agep16`). The table should look like the one shown below:

agep16	Freq.	Percent	Cum.
2	7	28.00	28.00
3	2	8.00	36.00
4	6	24.00	60.00
5	4	16.00	76.00
6	5	20.00	96.00
7	1	4.00	100.00
Total	25	100.00	

Now you can see that the cases have been assigned different values based on what their actual age is. If you want to double-check to make sure that this new variable contains the correct information, you could create a cross-tabulation of the original age variable by this new variable. Type `tab agecats agep16` in the Command window and press **Enter**. The following results table should be displayed:

```
(agecats_w |
   3) Age |
  variable |
 collapsed |
  into one |
      year |                      agep16
categories |        2         3        4        5        6        7 |      Total
-----------+------------------------------------------------------+----------
        18 |        7         0        0        0        0        0 |          7
        19 |        0         2        0        0        0        0 |          2
        20 |        0         0        6        0        0        0 |          6
        21 |        0         0        0        4        0        0 |          4
        22 |        0         0        0        0        5        0 |          5
        23 |        0         0        0        0        0        1 |          1
-----------+------------------------------------------------------+----------
     Total |        7         2        6        4        5        1 |         25
```

The table shows that you have created the new variable correctly. All 7 cases that are 18 years old are coded as being 2 years past 16 in the `agep16` variable. The 2 cases that are 19 are coded as 3 years past 16, and so on.

But you are not quite done. Remember the ultimate goal is to have a variable that represents the number of people the respondent has dated per year since turning 16. You need to divide the number of people each case reports to have dated by this newly generated variable that indicates how many years past 16 he or she is. To create this variable, follow the same steps as above. First, enter the command `-gen-`. Next, enter the new variable name you would like to use: `gen datpry16` (dated per year post 16). Then type an equal sign and put in the formula needed to correctly generate the new variable.

This latter portion is the slightly tricky part. But think of what information you need and where it is contained. You need the number of people each person reports to have dated (i.e., the `datnum` variable) and the number of years past 16 the person is (i.e., the previously created `agep16` variable). Then what mathematical operation do you need to perform to these pieces of information: Divide the former by the latter. Therefore, the full command that you need to enter into the Command window should read as `gen datpry16=datnum/agep16`.

Notice that the "/" is used to tell Stata to divide the value of each case on `datnum` by that case's value on `agep16`. See the "A Closer Look: Mathematical Operators and Their Symbols" box for a list of the most commonly used mathematical operators and their Stata symbols.

A Closer Look: Mathematical Operators and Their Symbols

When you create new variables in Stata, you often need to use a mathematical operation or function. Each of these operations has a particular symbol that Stata recognizes. Several are probably obvious, such as "+" for addition, but others might not be so clear. The following is list of the most commonly used operations/functions and their symbols.

Addition	+
Subtraction	–
Multiplication	*
Division	/
Power	^
Greater Than, Less Than	>, <
Greater Than or Equal to, Less Than or Equal To	>=, <=
Absolute Value	abs(number/variable name)
Natural Log	ln(number/variable name)
Square Root	sqrt(number/variable name)

Also note that Stata follows the traditional order of operations. For example, if for some reason you wanted the variable `datpry16` to contain the natural logarithm of the number of people dated per year after turning 16, you could type the following command:

```
gen lndatpry16=ln(datnum/agep16)
```

Stata will conduct what is inside the parentheses first (i.e., the division) and then take the natural logarithm of the result.

Just as before, it is generally a good idea to double-check that the new variable contains the information that you intended. Here, because the new variable was the result of an operation performed on two other variables, simply displaying a cross-tabulation will not provide all the information. Although there are several options to conduct this evaluation, perhaps the easiest method with the tools you have already learned is to open the Data Browser and check the values of a few cases on `datnum`, `agep16`, and `datpry16` to make sure that the calculation resulted in the correct information being coded in the new variable. Now that you know how to use the Command window, you can actually open the Data Browser by typing `browse` in the Command window and pressing **Enter**. Or you can type `browse datnum agep16 datpry16`, and only the variables listed will be displayed in the Data Browser window. Once you have opened the Data Browser, you should see that the new variable `datpry16` does indeed represent the number of people each case has dated per year since turning 16.

replace (if)

Sometimes, instead of creating a new variable, you might need to alter the values of an existing variable. There are several reasons why you would want to perform such an operation. You might realize that you need to make a replacement due to a mistake in the data entry process. For example, perhaps all the 18-year-olds are actually 19 years old. More frequently, you might want to combine categories for substantive reasons. For example, you might decide that you do not want to make the distinction between being out of the labor force and being out of work but looking for a job. Therefore, you would want to replace the values for the out of work but looking for a job cases to be equal to the value of those who are coded as unemployed so that these two categories would have the same value. In each of these scenarios, you would be asking your smart colleague to take the current values of that variable and replace them with a new value. In one word, you would ask your colleague to `-replace-` the old values with the new values.

In many respects, the `-replace-` command works very similarly to the `-gen-` command. Instead of creating a new variable, the `-replace-` command simply changes the values of a current variable. For the sake of practice, assume that the problem with age noted above had been discovered. Before starting the replacement process, it is helpful to produce a distribution of the existing variable to compare the new version of the variable after you replace the necessary values. Typing `tab agecats` in the Command window and pressing **Enter** produces the following results:

```
 (agecats_w3 |
      ) Age |
   variable |
  collapsed |
   into one |
       year |
 categories |      Freq.     Percent        Cum.
------------+-----------------------------------
         18 |          7       28.00       28.00
         19 |          2        8.00       36.00
         20 |          6       24.00       60.00
         21 |          4       16.00       76.00
         22 |          5       20.00       96.00
         23 |          1        4.00      100.00
------------+-----------------------------------
      Total |         25      100.00
```

You can see that there are 7 cases that need to have their value of 18 replaced to equal 19. Therefore, once you have made the replacement, the new version of `agecats` should have 9 cases that equal 19.

To fix this problem, you need to change all the cases that are currently coded as 18 to equal 19 on the `agecats` variable. As with previous commands, begin by typing the command into the Command window: `replace`. Next, you need to enter the name of the existing variable whose values you would like to replace: `replace agecats`. To keep this order in mind, it can be helpful to think about how you would ask a colleague to perform this operation (i.e., "please, replace (the values of the) agecats (variable)"). Now, just as with the `-gen-` command, you have to tell Stata what the new values should equal. In this case, you are asking Stata to set the new values to equal 19, so the command at this point should read `replace agecats=19`. Again, just as with `-gen-` what comes after the equal sign are usually numbers, mathematical operators, and variable names.

But you are not done at this point. The `-replace-` command, just like `-gen-`, performs whatever you enter after the equal sign in a case-by-case fashion on all cases, unless otherwise told not to. Therefore, if you pressed **Enter** with only the current command line typed, Stata would go through each case, and, because it has no further information, replace each respondent's value of `agecats` to equal 19. This change is not what you are trying to accomplish. Instead, you need to tell Stata that it should only change the value of a case *if* that case currently equals 18 on the `agecats` variable.

To perform this operation, you need to invoke an `-if-` statement. Whenever you use an `-if-` statment, you are telling Stata to only perform the command that precedes the `-if-`, on cases for which the expression that follows the `-if-` is true. Most often what will follow the `-if-` statement in a command line is a variable name with an equal to, greater than, or less than condition. In the current example, you want Stata to replace the value on `agecats` to equal 19 if the case's value on `agecats` equals 18. This command may be starting to sound a bit complicated but remember to think about how you would ask someone to complete this operation (i.e., "please, replace (the values of the) agecats (variable) (to) equal 19 if (the value of the) agecats (variable) equals 18"). Following this form leads to the correct structure of the command: `replace agecats=19 if agecats==18`.

But wait. You probably noticed that the equal sign after the `-if-` in the command line includes *two* equal signs. This is not a typo. When you want to use an equality expression in an `-if-` statement, you must type two equal signs. Again, this process may sound confusing, but there is a pretty straightforward rule of thumb to help keep this difference straight. Whenever you are telling Stata to make something equal to something else (i.e., to change a value so that it becomes equal to something else), then you only use one equal sign.

In the first part of this command, you are telling Stata to make the values on the variable `agecats` to become equal to 19, meaning you only need to type one equal sign. Whenever you are telling Stata to evaluate or assess whether something is equal to something else then you need two equal signs. In the second part of the above command, you are asking Stata to only perform the replacement after checking to see if that case's value on `agecats` is equal to 18, meaning you need to type two equal signs.

Type the `-replace-` command from above (replace agecats=19 if agecats==18) into the Command window and press **Enter**. Then produce a distribution table of the `agecats` variable (i.e., `tab agecats`). When you do you will see the following table:

```
 (agecats_w3 |
      ) Age |
   variable |
  collapsed |
   into one |
       year |
 categories |      Freq.     Percent        Cum.
------------+-----------------------------------
         19 |          9       36.00       36.00
         20 |          6       24.00       60.00
         21 |          4       16.00       76.00
         22 |          5       20.00       96.00
         23 |          1        4.00      100.00
------------+-----------------------------------
      Total |         25      100.00
```

The table shows that the replacement has been made correctly. All 7 cases that previously had been coded as 18 have now been set to equal 19. None of the values of the other cases have been altered, meaning the `-if-` statement operated successfully.

A Closer Look: The "Dreaded" Error Message

So far you have done everything correctly. Unfortunately, even the most experienced Stata users make mistakes. When you enter a command incorrectly, Stata displays an error message telling you that something went wrong. This error message is accompanied by a clickable link that displays more information about the particular error. Most frequently, error messages are the result of a typographical error (e.g., typing the name of a variable

(Continued)

(Continued)

incorrectly). Sometimes, however, error messages result from not entering the command or an option appropriately. One of the first such instances you probably will encounter is the use of the double equal sign after an `-if-` statement. For example, type the command you just conducted into the Command window, but delete one of the equal signs after the `-if-` portion of the command. The (incorrect) command should read: `replace agecats=19 if agecats=18`. When you press **Enter**, you will see the following results:

```
. replace agecats=19 if agecats=18
invalid syntax
r(198);
```

The error message is "invalid syntax," and "r(198)" is the clickable portion that tells you exactly why you received the error message. If you click on that portion of the message, Stata gives more information about the error. Note that at the end of this description, it states "Errors in specifying expressions often result in this message." Expressions include `-if-` statements, meaning Stata is telling you to double-check to make sure you have used the correct operator, which in this case you have not because you only entered one equal sign.

You may be realizing about now that there is a danger in using the `-replace-` command. Once you have performed a replacement, there is no reverting back. Now that you have changed the 7 cases that were previously recorded as 18 to equal 19, there is no way to separate them from the 2 cases that were originally coded as 19. So if you discovered that you had mistakenly thought those cases need their ages changed, it would be impossible to change them back. This situation is precisely why it is strongly recommended that you always work with a duplicate copy of the original data set and keep an original copy of the data file as a backup. Doing so will always provide you with a method for recovering the original version of any variable you may change. This scenario also illustrates the utility of "do files," explained in What Is a Do File? section of Chapter 3, as a method for saving all your commands, in case you need to replicate a portion of your analyses after making such a mistake.

Additionally, there is a safer way to use the -replace- command to prevent this scenario from occurring in the first place. For this example, assume that you have now learned that the one case that has been recorded as being 23 years old should have actually been coded as 22 years old. But you are concerned that at some point you may want to be able to identify this mis-coded case (perhaps to analyze why the case may have been incorrectly recorded). What you are looking to do then is create a second version of the agecats variable that you could replace the value of the 23-year-old case with 22. Before delving into how to create this copy, can you think of a way, using the commands you have already learned, to do so?

To create this copy, you need to use the -gen- command. You want to tell Stata to generate a new variable that is equal to the current agecats variable. Remember the structure of the -gen- command: command new-variable-name = new-value. Therefore, type gen agecatsrp=agecats in the Command window and press **Enter**. Now, create a cross-tabulation of the original agecats variable and your newly created copy (tab agecats agecatsrp). Doing so produces the following results:

```
(agecats_w  |
    3) Age  |
  variable  |
 collapsed  |
  into one  |
      year  |                         agecatsrp
categories  |        19         20         21         22         23 |     Total
-----------+-------------------------------------------------------+----------
        19  |         9          0          0          0          0 |         9
        20  |         0          6          0          0          0 |         6
        21  |         0          0          4          0          0 |         4
        22  |         0          0          0          5          0 |         5
        23  |         0          0          0          0          1 |         1
-----------+-------------------------------------------------------+----------
     Total  |         9          6          4          5          1 |        25
```

The two variables are identical, which is what you were looking to create. Now you can use the -replace- command just as you did before to change the value for the case coded as 23 to 22 on your newly created copy of the agecats variable. Remember, you are only asking Stata to replace a case to equal 22 if that case currently equals 23 on the agecats variable, meaning you need an -if- statement in your-replace-command. So you would type replace agecatsrp=22 if agecats==23 in the Command window and press **Enter**. Next produce a distribution table of the copied agecatsrp variable, now with the replaced value, and the original agecats variable (tab agecats agecatsrp):

```
(agecats_w |
    3) Age |
  variable |
 collapsed |
  into one |
      year |                   agecatsrp
categories |        19         20         21         22 |     Total
-----------+--------------------------------------------+----------
        19 |         9          0          0          0 |         9
        20 |         0          6          0          0 |         6
        21 |         0          0          4          0 |         4
        22 |         0          0          0          5 |         5
        23 |         0          0          0          1 |         1
-----------+--------------------------------------------+----------
     Total |         9          6          4          6 |        25
```

The values from the original agecats variable are listed in the rows and the new version's values are in the columns. The table shows that the two versions are similar, except that the one case that is coded as 23 in the original version has now been replaced to equal 22 on the new version. Notice that using this process allows you to be able to identify the case(s) that was replaced through a comparison of the old version of the variable with the copied and replaced version of the variable. This ability may be needed at some point in your analyses, and if nothing else you now will always have the original version of the variable in case you should need it.

Although this particular example may seem a bit superfluous, there are many more pertinent situations when you will need to replace the values of an existing variable. Even more frequently, you will find that this method of creating a new variable (either as the copy of an existing variable or based on a particular calculation) followed by a single or even multiple replacements to be quite useful.

One typical example of a more complicated replacement is the creation of adichotomous or nominal variable that is based on multiple conditions. Consider that you want to create a dichotomous variable (sometimes referred to as a dummy variable) that serves as an indicator of respondents being "isolated." For the present purposes, define "isolated" as having dated fewer than 2 people and having 2 or fewer friends. Based on this definition, you are going to count anyone who has dated 2 or fewer people *and* has 2 or fewer friends as isolated. As you can see, the conditions for being included are somewhat complicated. Fortunately, the combination of -gen- and -replace- help create this new variable.

To start, you need to create a new variable that can serve as the indicator of being isolated. To do so, you need to use the `-gen-` command, but remember you need to tell Stata what the new variable should equal. In the end, you want a dichotomous variable, meaning it should have two possible values. Typically, such variables are coded with one category equal to 0 and the other to 1. In the given example, the variable would be coded 0 if the case is not isolated and 1 if the case satisfies the requirements of being isolated.

A useful practice when creating these types of variables is to begin by coding every case on the new value to equal 0. Doing so essentially starts with the premise that no one is included in the indicator category. Then you can tell Stata which cases need to be replaced, to equal 1, if they satisfy the determined criteria to be included in the indicator. Thus to start, type `gen isol=0` into the Command window and press **Enter**.

Now you have the new variable in which every case has been set to equal 0. You need to replace the cases that fit the requirements for being isolated to equal 1. Notice, however, that based on the criteria there are two conditions that must be met, meaning you need to use a slightly more specific `-if-` statement. The `-replace-` command starts very similarly to what you did above: `replace isol=1 if`. From here you need to think about what you need to tell Stata so that it correctly replaces the cases to equal 1 if they meet all the requirements for being counted as isolated.

The first component is the case must have only dated 2 people or fewer. Therefore, the first clause in the `-if-` statement should clarify this condition: `replace isol=1 if datnum<=2`. Notice, instead of using a double equal sign as before, here you can use the less than or equal symbol because you are telling Stata to include all cases that are equal or less than 2 on `datnum`.

But you are not quite done. If you were to hit **Enter** at this point, all cases that have dated 2 or be fewer people would be replaced to equal 1. But you want cases to equal 1 (i.e., be in the isolate group) if they have dated 2 or fewer people *and* have 2 or fewer friends. Therefore, you need to add another condition to the `-if-` statement. Again, try to keep in mind what you would tell your smart colleague to do if you wanted them to actually move a group of people who should be counted as isolated. You might tell that person to "move everyone who has dated 2 or fewer people *and* who has 2 or fewer friends." If you replace each "has" in that statement with a less than or equal sign and the conjunctions with the appropriate symbols you will have the correct Stata command: `replace isol=1 if datnum<=2 & numfrien<=2`. Now you can press **Enter**. When you do so, you will see the following results:

```
. replace isol=1 if datnum<=2 & numfrien<=2
(2 real changes made)
```

These results tell you that 2 cases had their values changed on the variable `isol`. As before, it is good to double-check that the correct cases were replaced. You can use the Data Browser window to do so. Or you can produce a cross-tabulation of the `datnum` and `numfrien` to ensure that the correct cases were altered.

A Closer Look: Logical Operators and Their Symbols

When you use compound `-if-` statements, you need to use logical operators. Each of these operations has a particular symbol that Stata recognizes. The following is the list of most commonly used operators and their symbols.

And	&
Or	|
Not Equal To	!=

Also note that as with mathematical operators, it is helpful to use parentheses to ensure that the intended order of operators is followed. For example, if you wanted the cases to be coded as isolated if they are younger than 20 years old *and* have dated 2 or fewer people *or* have 2 friends or less, you *might* type

```
replace isol=1 if agecats<20 & datnum<=2 | numfrien<=2
```

Given this exact command, however, Stata would code people who are younger than 20 years and have dated 2 or fewer people as equal to 1. It would also code people who have 2 or fewer friends as equal to one. Note that this replacement is not what you intended because Stata grouped the first two clauses rather than the second two. To fix this scenario, use parentheses to explicitly denote the correct grouping:

```
replace isol=1 if agecats<20 & (datnum<=2 | numfrien<=2)
```

Now Stata will know that for a case to equal 1, it has to first be less than 20 years old and then the case either needs to have dated 2 or fewer people *or* have 2 or fewer friends.

To produce such a table, you want to create a standard cross-tabulation as you did above. Except now you only want to display the cases that were replaced to equal 1 on the new variable so that you can check to make sure those cases have the appropriate values on datnum and numfrien. In other words, you only want the -tab- command to apply to cases that are now equal to 1 on the isol variable. To limit the table to only these cases, you should invoke another -if- statement. Type tab datnum numfrien if isol==1 in the Command window and press **Enter**. Remember because you are asking Stata to assess whether a case is coded as 1 on the isol variable (i.e., comparing each case's value to 1), you need the double equal sign.

```
                | (numfriend
                | _w3) N:1.
                |  Now for
                |  the next
(datnum_w3) [IF |   set of
 HAS BEEN IN A  | questions,
      ROMANTIC  |  I'll be
RELATIONSHIP OR |   asking
      HAS BEEN  |    some
  MARRIED J:4.  |   things
           How  |          2 |     Total
----------------+------------+----------
             1  |          1 |         1
             2  |          1 |         1
----------------+------------+----------
         Total  |          2 |         2
```

The table only shows the values on datnum and numfrien for the two cases that were changed to be equal to 1 on the isol variable. And as you can see, they both satisfy the requirements of having dated 2 or fewer people and having 2 or fewer friends. Using this type of -if- statement with other commands is covered in later chapters that examine specific analyses. For now, you should be able to use it to assess whether these type of replacements were completed successfully.

recode

One of the limitations of the -replace- command is that it can only replace one value at a time. That is, in the agecats example above, you needed to type two commands to change the values of 18 to 19 and 23 to 22. You cannot type -replace-, in one command line, to make agecats equal to both 19 and 23 based on different conditions. In some respects, this limitation could be seen as a strength of the -replace- command because it makes you be very careful and explicit when you make changes to the data. But it can also become quite cumbersome when you need to alter several values of an existing variable. Fortunately, Stata has a command, `-recode-`, that allows multiple replacements using only one command line.

Intuitively, you may not arrive at the word "recode" for a command that asks a colleague to change the values of a variable. But the command name provides a good deal of intuitive information that should help you remember it. The primary distinction between `-replace-` and `-recode-` is that with `-replace-` you are literally overwriting the value of a variable (or a case) with another value. Although you used an `-if-` statement to isolate this overwriting to particular cases, the default operation of `-replace-` is to change the values of every case in a variable. `-recode-`, on the other hand, is asking Stata to take a current value and change it (i.e., recode it) into another value. This different operation means you have to tell Stata the old value you would like to change and what new value you would like it to be recoded to.

For this example, consider that you have decided to create an ordinal variable out of the currently interval-ratio variable `datnum`. You want to distinguish minimal daters, moderate daters, and frequent daters. Therefore, you need to recode the current 13 categories into 3. To make the three distinctions, it is best to begin by looking at a distribution of the variable. Type `tab datnum` into the Command window and press **Enter**.

```
. tab datnum

(datnum_w3) [IF |
  HAS BEEN IN A |
       ROMANTIC |
RELATIONSHIP OR |
       HAS BEEN |
   MARRIED J:4. |
            How |      Freq.     Percent        Cum.
----------------+-----------------------------------
              1 |          1        4.00        4.00
              2 |          4       16.00       20.00
              3 |          4       16.00       36.00
              4 |          1        4.00       40.00
              5 |          3       12.00       52.00
              6 |          2        8.00       60.00
              7 |          1        4.00       64.00
              9 |          1        4.00       68.00
             10 |          2        8.00       76.00
             11 |          1        4.00       80.00
             15 |          2        8.00       88.00
             20 |          3       12.00      100.00
----------------+-----------------------------------
          Total |         25      100.00
```

Based on these results, you could decide to classify anyone who has dated 3 or fewer people as minimal daters, 4 to 7 as moderate daters, and anything more than 7 as frequent daters. You could use the steps shown above to create a new variable and then replace its values using appropriate if statements. Although this method would produce the desired result, it would require four separate commands (one to create the new variable, and one for each of the three categories).

To simplify this process, you can use the `-recode-` command. The structure of the `-recode-` command is similar to all the ones you have been working with in this chapter. Start by typing the command name and the original variable whose values you are changing: `recode datnum`. Now you need to tell Stata which values to recode and what new values they should take. You could choose any three numbers to represent the three new categories, but it is generally easiest to use 1, 2, and 3 to code these types of ordinal variables. You will be asking Stata to recode any cases who are currently coded as having dated 1 through 3 people to now equal 1 (minimal daters), currently coded as having dated 4 through 7 people to now equal 2 (moderate daters), and anyone who has dated 7 through 20 people should be changed to equal 3 (frequent daters). The `-recode-` statement, therefore, requires three separate conditions be specified.

The command you need to type in the Command window to accomplish this alteration is: `recode datnum (1/3=1) (4/7=2) (8/20=3)`. Before you press **Enter**, take a moment to examine the command. The first new symbol you probably notice is the backslash "/." This is the operator for "through," which tells Stata to change all values from and including the number listed before the "/" through and including the value listed after the "/" . It might be confusing that the "/" operator is used both to symbolize the mathematical function of division and this logical function of "through." Fortunately, the `-recode-` command is virtually the only situation in which "/" is used to signify "through". In basically every other possible scenario, "/" is a symbol for division.

Next, you see that you only need to use a single equal sign. Here you are telling Stata to set the values of the cases on a particular variable to be equal to something else, which means you only need the one equal sign. Finally, if you look at the distribution of the `datnum` variable shown above, you might notice that there are no cases who have dated 8 people. The final condition in the `-recode-` command tells Stata to recode all values from and including 8 through and including 20 to equal 3. It does not matter that none of cases in the current data are actually coded as 8, there also are not any cases coded as 13 or 18. Stata is simply looking at the range, and any case that has a value that falls in that range will be recoded.

Now you may be ready to press **Enter**, but pause for just another moment. Think about what you are asking Stata to do. Stata will be changing all the values of the `datnum` variable to be equal to 1, 2, or 3. There might be a situation later in your analyses where you want to know the exact number of people each case has dated. If you continue with the -recode- command as written, you would have no way to return to the original values (short of reopening the original data file).

You already know one way to protect against this danger. Just as with the -replace- command, you could create a new variable that is the exact copy of the datnum variable and then recode this duplicated version. This method would definitely work, but it would require an additional step. Fortunately, the -recode- command has a useful option to protect against this danger that does not require an additional command. You can invoke the -generate(newvar)- option to automatically create a new variable that contains the recoded version of the original variable. This option is slightly different from the ones you have worked with before in that it requires you to type not only the option but also something else in the parentheses. The "newvar" portion of the option denotes that you need to type a name for the new variable that will be created and hold the recoded values.

Putting everything together, keep what you have already typed and add the generate option into the Command window, recode datnum (1/3=1) (4/7=2) (8/20=3), gen(datlevs), and press Enter. Now you have told Stata the values on datnum that need to be recoded, the values they should take, to create a new variable to hold these recoded values, and the name of that new variable (datlevs). As always, it is useful to check to make sure the command recoded everything in the way you intended. Type tab datnum datlevs in the Command window and press **Enter**. The following cross-tabulation is presented:

```
(datnum_w3) [IF |
  HAS BEEN IN A |
       ROMANTIC |
RELATIONSHIP OR |   RECODE of datnum ((datnum_w3)
       HAS BEEN |      [IF HAS BEEN IN A ROMANTIC
   MARRIED J:4. |       RELATIONSHIP OR HAS BEEN MARRIED]
            How |          1          2          3 |     Total
----------------+---------------------------------+----------
              1 |          1          0          0 |         1
              2 |          4          0          0 |         4
              3 |          4          0          0 |         4
              4 |          0          1          0 |         1
              5 |          0          3          0 |         3
              6 |          0          2          0 |         2
              7 |          0          1          0 |         1
              9 |          0          0          1 |         1
             10 |          0          0          2 |         2
             11 |          0          0          1 |         1
             15 |          0          0          2 |         2
             20 |          0          0          3 |         3
----------------+---------------------------------+----------
          Total |          9          7          9 |        25
```

You can see from this table that the cases from the original `datnum` variable were recoded into the correct values on the new `datlevs` values. For example, the 1 case that reported having dated 1 person, the 4 that had dated 2 people, and the 4 that had dated 3 people are now all equal to 1 (minimal daters) on the new `datlevs` variable. Additionally, Stata automatically has added some information to the `datlevs` variable, shown at the top of the table, to help you remember that it is the recoded version of the `datnum` variable.

As you work with quantitative data, whether it is a secondary data set or one that you have collected yourself, one of the most frequent alterations you will need and/or want to make is this type of recategorization. Surveys often ask for more precise responses than you may need. Therefore, being able to collapse multiple categories into broader groupings is one of the most vital skills in an analyst's repertoire. The `-recode-` command is the most effective command for making this type of change.

A Closer Look: Multiple Commands to the Same Ends

You may have noticed that the `-replace-` and `-recode-` commands seem similar and that you could use each to achieve the same goal. You will notice several other instances throughout this book in which multiple pathways, using various commands, lead to the same outcome. This equifinality may at first seem confusing and may even lead to some frustration by making it seem as if you have to remember when to use which command.

But viewed from another perspective, this feature is one of Stata's greatest strengths—its flexibility and adaptability. You do not need to worry about which combination of commands is the "right" way, as long as the combination you use produces the desired end. Some users may prefer, for a variety of reasons, to use a combination of `-gen-` and `-replace-` rather than `-recode-` with the `-gen(newvar)-` option, whereas others may even prefer to use a combination of `-gen-` and then `-recode-`. All three patterns would create a similar variable, so you could use whichever command combination that seems most straightforward, comfortable, or even easiest to remember. In these types of situations, there is no one correct way to accomplish the task. Stata allows users to construct the pattern that works for them. So do not try to remember each and every possible method, rather practice the one in which you have the most confidence. As long as the outcome is what you are looking for, then that is the "right" way.

Nonessential, Everyday Commands

The 5 commands that have been covered to this point are the most essential commands both in their utility and in their applicability to learning the foundation of Stata. There are, however, a handful of other commands that are frequently used, although perhaps not explicitly essential to the completion of a research project. These commands do not "fit" neatly into any one substantive section, which is partly why they have been included in this somewhat miscellaneous subsection. Most of the commands are fairly straightforward and therefore are described relatively briefly. Still the details provided explain how and when to use them effectively.

rename

When you use secondary data sets, the variable names may not always be extremely informative. There are data sets that even name variables based on the survey question (e.g., `q112`) making it difficult to remember what information is stored in the variable.

Chapter 1 explained two ways to alter the name of variables using the point-and-click interface. The command to do so is similarly straightforward, making all three efficient alternatives. The command name is `-rename-` followed by the existing variable name and the new name you want it to be called.

For example, in the current data it may be helpful to have the `gender` variable named `female` so that it is clear that the variable is an indicator of being female (i.e., female is coded as 1).

Type `rename gender female` into the Command window and press **Enter**. The variable `female`, in place of `gender`, now appears in the Variables window.

drop/keep (if)

Although it is generally not necessary, there are times when you might want to eliminate a variable or particular cases from a data set. The command to accomplish both tasks is `-drop-`, but the structure to accomplish either task is slightly different.

To drop a variable or variables from the data set, simply enter the command followed by the names of the variables to be deleted. In the current data set, the `examp` and `examp2` variables were generated only as illustrations. Because you do not need them for your analyses, you could choose to eliminate them from the data set.

Type drop examp examp2 in the Command window and press **Enter**. The two variables are now eliminated from the data file and no longer appear in the Variables window.

When you need to eliminate particular cases, think about how you might ask your smart colleague to do so. For instance, you might realize that the one 23-year-old case was included in the data set by mistake and should be completely eliminated. You might ask your smart colleague to "drop cases if they are 23 years old." This verbal command is similar to the Stata command to drop cases. You must invoke an -if- statement immediately after the -drop- command.

Type drop if agecats==23 in the Command window and press **Enter**. Remember that the double equal sign is needed in the -if- clause because you are asking Stata to assess whether a statement is true. If you now produce a distribution table of the agecats variable, by typing tab agecats into the Command window and pressing **Enter**, the table displays as

```
(agecats_w3 |
  )     Age |
   variable |
  collapsed |
   into one |
       year |
 categories |      Freq.     Percent        Cum.
------------+-----------------------------------
         19 |          9       37.50       37.50
         20 |          6       25.00       62.50
         21 |          4       16.67       79.17
         22 |          5       20.83      100.00
------------+-----------------------------------
      Total |         24      100.00
```

The variable now only ranges from 19 to 22, and the total number of cases is 24 instead of 25.

The -keep- command is implemented exactly like the -drop- command, but produces the opposite result of *retaining* only the cases or variables specified keep in the command line. For example, you might decide you only want to keep respondents who are younger than 21 years of age. Now instead of specifying who to eliminate, using the -if- statement, you need to tell Stata who to keep. Because you only want to keep cases who are younger than 21 years of age, you would type keep if agecats<21 into the Command window and press **Enter**. After doing so, typing `tab agecats` into the Command window and pressing **Enter** displays the following table:

```
(agecats_w3 |
     ) Age |
  variable |
 collapsed |
  into one |
      year |
categories |      Freq.     Percent        Cum.
-----------+-----------------------------------
        19 |          9       60.00       60.00
        20 |          6       40.00      100.00
-----------+-----------------------------------
     Total |         15      100.00
```

The variable now only ranges from 19 to 20, and the total number of cases is 15 instead of 24.

One note of caution to keep in mind when deciding whether to drop (or keep) selected cases from a data set. It may seem appropriate to completely eliminate cases that do not pertain to particular analyses. For instance, you may only be interested in the number of friends among young adult females. If so, you might consider dropping all the males from the sample. There is no technical problem with doing so, but it is not the most effective approach. You could of course keep a backup data file that contains the full sample. But all the data manipulations you make on your restricted data would then need to be redone if your research interests change, and you decide to include males in the analyses. It is generally a better strategy to use an `-if-` statement with your desired analyses. In the above example, if you wanted the cross-tabulation shown in this chapter to only include people who were younger than 23 years of age, you could type `tab employst gender if agecats<23`. This latter command would accomplish the same ends as if you dropped the case and then produced the cross-tabulation, but it would not permanently alter the data set.

describe

The `-describe-` command (shortened `-desc-`) can be used to present information about the entire data set or about particular variables. It can be entered completely by itself, which will display detailed information about the data file and each variable included in the data set. The inclusion of the Properties window in Stata 12 makes this a less needed command, but for users of Stata 11 or older, it is still very helpful.

Type `desc` into the Command window and press **Enter** to see this full set of results. Invoking the `-short-` option provides only the information about the data. Type `desc, short` into the Command window and press **Enter**. The following, more concise results are displayed:

```
Contains data from C:\Documents and Settings\klongest\My
Documents\Stata\Data\Chapter 2\Chapter 2 Data.dta

 obs:          15

 vars:         12

 size:        915

 Sorted by:

 Note: dataset has changed since last saved
```

The results indicate how many cases (`obs`) are in the date file, how many variables (`vars`) are included in the data, and the size of the data file.

If you want information about only particular variables, you can enter their names after the `-desc-` command. Type `desc female datlevs` into the Command window and press **Enter**. The information shown below is presented:

```
variable name  type  format   label   variable label
--------------------------------------------------------------
female      byte  %12.0f   gender   (gender_w3)Respondent gender
datlevs      int  %9.0g        RECODE of datnum ((datnum_w3)
                              [IF HAS BEEN IN A ROMANTIC RELATIONSHIP
                              OR HAS BEEN MARRIED]
```

Several aspects of the variables are shown, including the type and format (both discussed in Chapter 1). Additionally, this display lists the value label (`label`) attached to the variable and its variable label. Both of these aspects of variables are discussed in the Data Management: Working With Labels section of Chapter 3. But for now it is useful to know that the `-desc-` is a useful command to obtain a quick summary of all this information.

display

One interesting side utility of the Command window is that it can serve as a calculator. When you execute the `-display-` (shortened `-disp-` or even `-di-`) command followed by a mathematical formula, Stata will display the answer in the Results window.

For instance, if you wanted to double-check the percentage of males who are employed, displayed in the `employst` by `gender` cross-tabulation, you can type `di (2/9)*100` in the Command window and press **Enter**. The correct result, 22.22, appears in the Results window.

The `-di-` command follows a standard mathematical order of operations and uses the same symbols for operators discussed in the "A Closer Look: Mathematical Operators and Their Symbols" box. Whenever you enter numbers that are four digits or more, the comma(s) should not be typed. If you wanted to know the square root of 50,345, you would type `di sqrt(50345)` into the Command window, *not* `(50,435)`.

Although it may not be the most vital command, the `-di-` command is nonetheless extremely useful and just one more positive feature of Stata.

set more off

When you run a command that produces lengthy results (e.g., a frequency distribution of a variable with numerous categories), the Results window is forced to scroll. Stata, however, by default prevents the results from scrolling past one page view. When this situation arises, Stata pauses the display of the results and displays the word `more` at the bottom of the screen. Essentially, Stata is asking you whether you would like to see "more" of the results. To see the rest of the results, you can either click on the displayed `more` with your mouse or you can press any button on your keyboard. At first, this default operation may seem a bit frustrating, but Stata is actually saving you the work of scrolling back up to see the first set of results by giving you a chance to review them before continuing.

If you prefer to have Stata simply show all the results without pausing, you can enter the command `-set more off-` in the Command window and press **Enter**. Furthermore, if you would like to change the default setting so that Stata never pauses while displaying the results, you can invoke the `-permanently-` command along with the `-set more off-` command. If you would ever like to revert to the original default and have Stata pause when displaying lengthy results, you can invoke the `-set more on-` command.

Summary of Commands Used in This Chapter

```
*5 Essential Commands
*tab
tab employst
tab employst, sort
tab employst gender
```

```
*sum
sum datnum
sum datnum agecats
sum datnum, detail

*generate
gen examp=200
tab examp
gen examp2=21-16
tab examp2
gen agep16=agecats-16
tab agep16
tab agecats agep16
gen datpry16=datnum/agep16

*replace if
tab agecats
replace agecats=19 if agecats==18
tab agecats
gen agecatsrp=agecats
tab agecats agecatsrp
replace agecatsrp=22 if agecats==23
tab agecats agecatsrp
gen isol=0
replace isol=1 if datnum<=2 & numfrien<=2
tab datnum numfrien if isol==1

*recode
tab datnum
recode datnum (1/3=1) (4/7=2) (8/20=3), gen(datlevs)

*Nonessential, Everyday Commands
*rename
rename gender female

*drop
drop examp examp2
drop if agecats==23
tab agecats
keep if agecats<21
tab agecats
```

```
*desc
desc
desc, short
desc female datlevs

*display
di (2/9)*100
di sqrt(50345)

*set more off
set more off
set more off, permenantly
set more on

*A Closer Look: The Dreaded Error Message
replace agecats=19 if agecats=18
```

Exercises

Use the original `Chapter 2 Data.dta` for the following problems.

1. Produce a distribution table for the `numfrien` variable.
2. Produce a distribution table for the `numfrien` variable that is sorted by frequency.
3. Create a cross-tabulation of `numfrien` and `agecats`.
4. Reproduce the cross-tabulation of `numfrien` and `agecats` so that the percentages of each age category are displayed.
5. Generate a new variable that is the sum of the number of friends (`numfrien`) and number of people each respondent has dated (`datnum`).
6. Generate a new variable that replaces all respondents who report dating (`datnum`) 20 people to be equal to 15.
7. Generate a new variable that is an indicator of people who are 20 years old (`agecats`) and have dated (`datnum`) between 5 and 10 people.
8. Recode, and create a new variable containing the recoded values, the `agecats` variable into a two-category variable representing the respondents being younger than 21 years old versus being 21 years of age or older.
9. Rename the `numfrien` variable to be called `frndnum`.
10. Using the Command window, calculate the product of 976 and 543.

3

Do Files and Data Management

Now that you feel confident with the most essential and frequently used commands, you may be ready to start performing statistical analyses. Before jumping into the strategies for analyzing data, it is important to cover some of the foundational principles of working with data. Although not as exciting as estimating means and regression coefficients, these data management techniques are essential to becoming a proficient Stata user and effective quantitative analyst. It may be tempting to skip this chapter in order to learn the pragmatic commands that will help you complete an assignment. If that is your only need for Stata, then you should by all means go directly to the next chapter. But if you have any intention of using Stata, even if only for a single research project, the topics covered in this chapter will save you hours of time and frustration, both in the short and in the long term. Developing effective data management habits when you are first beginning to use Stata is much easier than breaking inefficient habits after you have been using the programming for an extended period.

This chapter begins by explaining perhaps one of the most useful components of Stata—the "do file." A do file is a way to save all of the commands you conduct during a particular project for later use, alteration, or replication. As you are just now becoming more comfortable using the Command window to perform operations, it may seem intimidating to learn a new method for conducting these tasks. Do files function very similarly to the Command window, and with a little practice you will be just as adept at using both. You will find that the ability to use do files effectively not only minimizes aggravation but exponentially increases your productivity. Finally, the chapter examines several data management "best practices," including working with labels, missing data, string variables, and saving results.

All the examples that follow use the `Chapter 3 Data.dta`, available at **www.sagepub.com/longest**. This data set contains the same 7 variables from the National Study of Youth and Religion (NSYR) data that were used in Chapter 2, but the Chapter 3 data set contains the full Wave 3 sample of 2,532 young adults. As mentioned in Chapter 1, it is a good idea to save a copy of the data file you are working with so that you always have a backup of the original data.

What Is a Do File?

While you were working through the previous chapter, you may have wondered what happened to the commands once they were run. You might have questioned what would happen if you ever wanted to generate that same variable or make similar recodes again. You saw how you could retrieve commands during a single Stata session, but once Stata is closed, the Review window and command history are eliminated. If you had to complete the previous chapter during two different sessions, you probably realized that when you started your second session all those data changes and analyses you conducted during the first session were nowhere to be found. This inability to save commands might have even caused a reasonable amount of irritation and/or led you to incessantly save your data. The latter method, while frequently used, has several disadvantages, which are discussed below. These drawbacks notwithstanding, saving numerous copies of the data does not save any of the analyses (e.g., summary statistics or cross-tabulations) that you performed. If you ever needed to recreate a particular analysis and you had been conducting them solely through the Command window, you would have to retype in all the necessary commands. This replication is common in statistical analysis, and clearly, the method of always retyping every single command is inefficient.

Fortunately, Stata has a built-in mechanism for addressing this challenge: the "do file." A do file is so named because it "does" a set of commands saved within it, and its file extension (i.e., the part of the file name after the ".") is .do. Although they might seem like a completely new part of Stata, do files are extremely similar to long Command windows that you can save. Their operation is similar to a word processing file, making their interface very familiar and easy to navigate.

The primary purpose of a do file is to allow you to save all the work that you have done. Do files are a place to keep a record of all of the commands that you have used in conducting a particular research project. For example, you could have used a do file to save all the commands you learned and used in Chapter 2 to both change the data and conduct statistical analyses on that data. Do files, therefore, are essential for completing analytic projects that require more than one Stata session.

If you are taking your first steps into the data analysis world, this utility may not seem important. Perhaps a brief example will further emphasize the necessity of do files. Imagine that the analyses you conducted in Chapter 2 were for an actual research project that was investigating what types of young adults belonged to the various employment/school combinations. In Chapter 2, you produced a cross-tabulation of gender by employment status, which also contained the column percentages. But what if you realized that you also needed the row percentages and the chi-square statistic? You would have to go back and retype the full command line with the new, added options. Whereas if the command had been saved in a do file, you would only have to type the two new options.

Perhaps this time saver still does not convince of the need for learning how to use do files. Consider, however, that you would like to see whether young adults who are classified as "isolated" are more likely to be out of the labor force than young adults who have friends or romantic relationships. In Chapter 2, you created a variable to indicate being isolated by categorizing all young adults who had dated 2 or fewer people and who had 2 or fewer friends as isolated. Now perhaps you saved a version of the data that contains this newly created variable and can relatively easily open that data set to conduct the necessary cross-tabulation. What if you realize, however, that your definition of isolated is too weak and decide that young adults should only be categorized as isolated if they have dated 1 or fewer people and have 1 or fewer friends. Now you would be forced to completely retype out two full command lines to create this new version of the `isol` variable. As you will learn next, if you had created the original `isol` variable in a do file, you could simply alter the "2s" in the appropriate command lines to "1s," and the `isol` variable would be created according to your new definition.

As you might imagine, these types of replications with alterations can be quite common in quantitative research projects. Having to retype these commands might seem to be only a minor nuisance, but when you are conducting full research projects the number of commands needed can become quite large very quickly. Meaning these minor nuisances can pile up and become a major aggravation. Not to fear, all this stress can be avoided by learning the basics of using the do files.

OPENING AND SAVING DO FILES

There are three ways to open the Do File window. First, select the **Window** menu, then choose **Do-File Editor**, and select **New Do-File Editor**. Second, you can hold the **Ctrl** button on your keyboard and hit the number **9**. Finally, you can simply click on the icon, which is next to the **Data Editor** button. Once you have completed one of these methods, your screen should look like this:

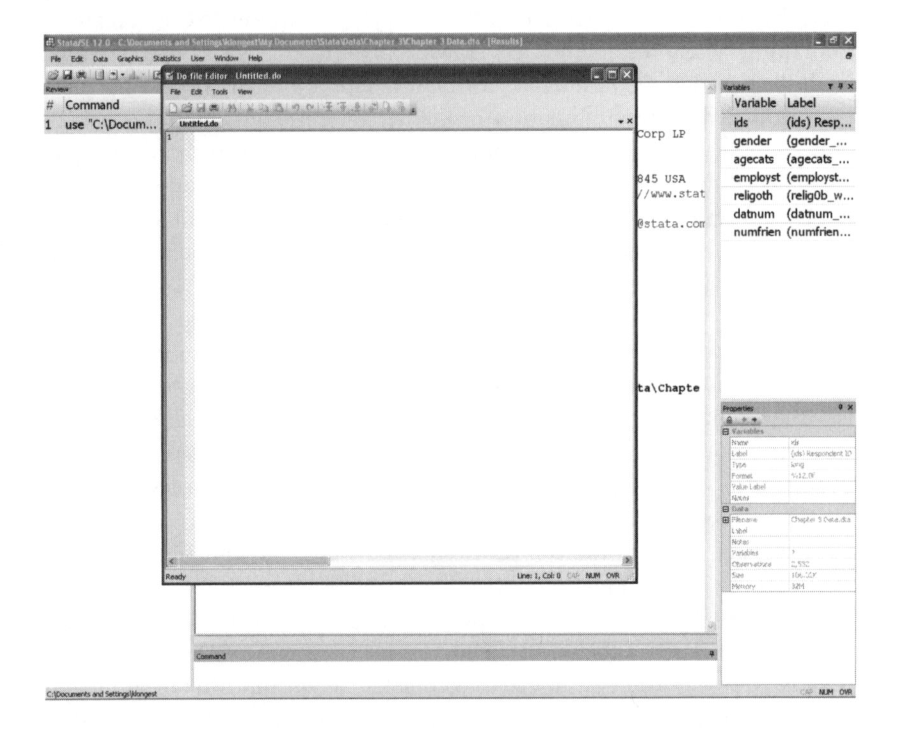

The new Do File window now appears. You can resize, move, or minimize the Do File window as you would with any window. You can see that when you open a new do file, Stata gives it the default name of `Untitled.do`.

Before you start entering commands into the do file, it is a good idea to save the do file. Saving a do file is very similar to saving a data set, except you use the menus within the Do File window. On the Do File window, open the **File** menu and click on **Save As**. You can then enter a new file name, such as `Chapter 3 Commands.do`, and click **Save**. To reopen this saved do file during a later Stata session, follow the steps from above to open the Do File window. Then you can either use the **File** menu, by selecting **Open** after clicking on **File**, or the icon in the Do File window to find and open an existing do file.

TRANSLATION FROM THE COMMAND WINDOW

Now that you have a saved Do File open, you can begin entering commands into it. It may be a bit daunting, however, to start inputting commands from scratch. An effective method to begin with is to copy the commands that you run in the Command window into your do file.

For example, in the fictive research example given above, you were interested in the employment/education patterns of young adults. To start this investigation, you might want to examine the gender differences in young adults' employment and schooling combination. In Chapter 2, you learned how to conduct a basic cross-tabulation that would help see if there were such differences. First, select the Command window. Note that when you do so, the Do File window is automatically hidden behind the other Stata windows. Next, type the appropriate command: `tab employst gender, col` in the Command window and press **Enter**. Remember the options `-col-` is asking Stata to produce the percentages within the columns. When you do, the following results should be displayed:

```
. tab employst gender, col

+-------------------+
| Key               |
|-------------------|
|     frequency     |
| column percentage |
+-------------------+

                     | (gender_w3)Respondent
     (employstat_w3) |        gender
   Employment Status |      Male      Female |     Total
---------------------+-------------------------+-----------
   Out of labor force |        27          58 |        85
                     |      2.19        4.46 |      3.36
---------------------+-------------------------+-----------
No school or work but |        65          62 |       127
                     |      5.28        4.77 |      5.02
---------------------+-------------------------+-----------
            Employed |       375         325 |       700
                     |     30.44       25.00 |     27.65
---------------------+-------------------------+-----------
  Employed and school |       404         547 |       951
                     |     32.79       42.08 |     37.56
---------------------+-------------------------+-----------
       In school only |       306         292 |       598
                     |     24.84       22.46 |     23.62
---------------------+-------------------------+-----------
  Active armed forces |        53          14 |        67
                     |      4.30        1.08 |      2.65
---------------------+-------------------------+-----------
      Legitimate skip |         2           2 |         4
                     |      0.16        0.15 |      0.16
---------------------+-------------------------+-----------
               Total |     1,232       1,300 |     2,532
                     |    100.00      100.00 |    100.00
```

This is the type of analysis that you might want to have saved in case you ever need to replicate it or rerun it with a slight alteration. To move this command into your do file, you can copy and paste the command from the Results window, which is listed just above the cross-tabulation key. To do so, highlight the command (`tab employst gender, col`) using your mouse and then either click **Ctrl+C** or right click and select **Copy**. Next, you need to bring up the Do File window by either selecting it from the Taskbar at the bottom of your screen or by clicking on the small arrow next to the icon and selecting the do file. Once you have the do file window selected, paste the command, either by pressing **Ctrl+V** or right click and select **Paste**, into the do file. When you do so, the screen should appear as follows:

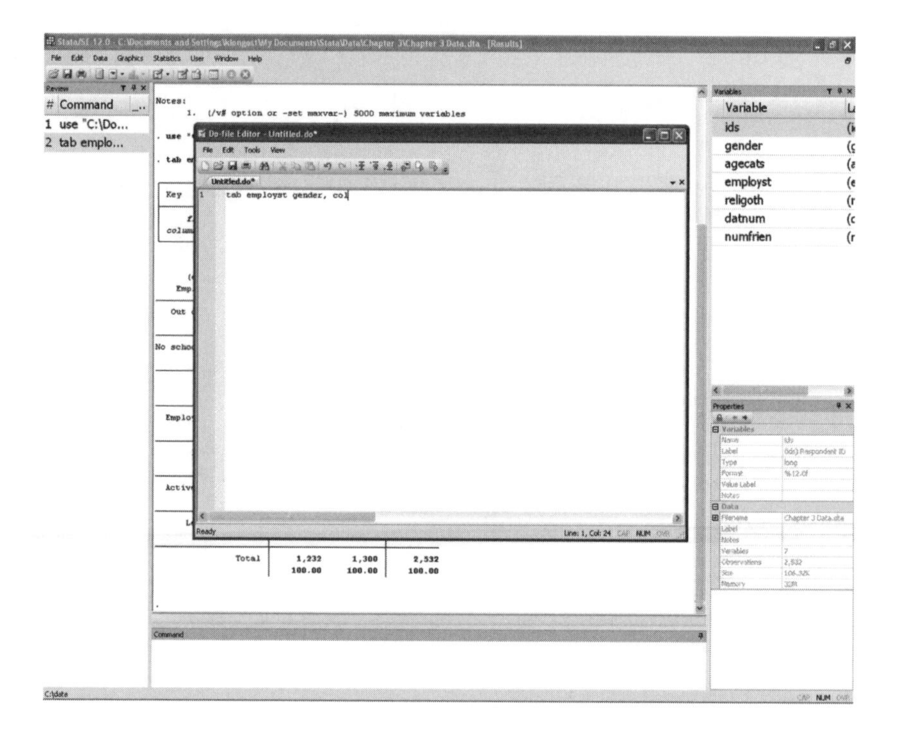

Now the command to produce the cross-tabulation will be saved in your do file. One word of caution when using the above method. When you copied and pasted the command from the Results window, you may have copied a "." at the beginning of the command line. This "." should not be included in your do file. The "." is used in the Results window as a way to distinguish a command that already has been run. You are using your do file to run these commands in the future, meaning you do not want to include the "." in the command. If you did paste it into your do file, simply delete it. Another method that accomplishes the same goal but helps prevent this potential problem is to copy the command from the Command window. Start by selecting the Command window and pressing **Page-Up** to display the previously run command. Then select, copy, and paste the command into your do file just as you did above.

You may have noticed, if you are using Stata 11 or 12, that the do file name now has an "*" listed after it both at the top of the Do File window and on the particular do file tab. This "*" indicates that your do file has been altered since the last time you saved it. You should save do files similar to how you save other documents. You probably do not need to save it after every single alteration, but it is a good practice to save your do file frequently. A shortcut to do so is by simply clicking the icon.

Now that you have a command in your do file, you can execute it without having to retype it into the Command window. For example, imagine that you had ended the Stata session where you produced the employment status by gender cross-tabulation and then realized that you did not move this cross-tabulation into your research report. (Note: Methods for saving the results are discussed in more detail in the Saving Results section at the end of this chapter.). Because you used a do file, you do not have to retype the command to produce the desired cross-tabulation. Rather you can simply press the icon located in the middle of the top of the Do File window. Once you have pressed this button, you should see that the cross-tabulation results have been redisplayed in the Results window. Notice that this cross-tabulation is identical to the one produced when you executed the command using the Command window. The results, however, now include a line that looks something like

```
. do "C:\DOCUME~1\klongest\LOCALS~1\Temp\STD0b000000.tmp",
```

which indicates that the command was executed using a do file.

Using the copy and paste method as a way to store the commands you have run is an effective way to become more comfortable with do files. The real payoff for using the do files, however, comes when you use them to enter new commands. For example, in your analyses of young adults' employment status, you may realize that age is an important factor. Therefore, you decide to look at a cross-tabulation of `employst` by `agecats`. You could, of course, enter this command into the Command window, press **Enter**, and then copy and paste the command into the do file. Once you feel confident with the basic operation of do files, however, it is more efficient to type the command directly into the do file and execute it from there.

The main difference between do files and the Command window is that do files do not recognize **Enter** as a way to execute a command. Remember, to run the cross-tabulation from the do file, you had to press the icon, not **Enter** as you would have had you inputted the command in the Command window. In fact, a do file thinks that every new line (i.e., every time **Enter** is used) is a new command, which is why you do not need to include any sort of symbol to denote the end of a command line (including a carriage return [i.e., pressing **Enter**] inherently tells Stata that the command line is finished). Therefore, to produce your new cross-tabulation, you first need to press **Enter** to move to a new line. Once you are on the new line, type in the necessary command just as you would if you were entering it in the Command window: `tab employst agecats, col`. To execute this command, press the icon. The following results are displayed:

```
. tab employst agecats, col

+-------------------+
| Key               |
|-------------------|
|     frequency     |
| column percentage |
+-------------------+

     (employstat_w3) | (agecats_w3) Age variable collapsed into one year categories
   Employment Status |        17         18         19         20         21         22         23         24 |     Total
---------------------+----------------------------------------------------------------------------------------+----------
  Out of labor force |         0          9         14         20         15         24          3          0 |        85
                     |      0.00       1.87       2.79       3.69       2.90       5.77       4.76       0.00 |      3.36
---------------------+----------------------------------------------------------------------------------------+----------
No school or work but |        0         22         31         23         28         18          5          0 |       127
                     |      0.00       4.57       6.19       4.24       5.42       4.33       7.94       0.00 |      5.02
---------------------+----------------------------------------------------------------------------------------+----------
            Employed |         1         82        113        158        163        152         31          0 |       700
                     |      9.09      17.05      22.55      29.15      31.53      36.54       9.21       0.00 |     27.65
---------------------+----------------------------------------------------------------------------------------+----------
 Employed and school |         2        177        199        212        211        132         17          1 |       951
                     |     18.18      36.80      39.72      39.11      40.81      31.73      26.98     100.00 |     37.56
---------------------+----------------------------------------------------------------------------------------+----------
      In school only |         8        179        133        116         83         72          7          0 |       598
                     |     72.73      37.21      26.55      21.40      16.05      17.31      11.11       0.00 |     23.62
---------------------+----------------------------------------------------------------------------------------+----------
 Active armed forces |         0         12         10         12         16         17          0          0 |        67
                     |      0.00       2.49       2.00       2.21       3.09       4.09       0.00       0.00 |      2.65
---------------------+----------------------------------------------------------------------------------------+----------
     Legitimate skip |         0          0          1          1          1          1          0          0 |         4
                     |      0.00       0.00       0.20       0.18       0.19       0.24       0.00       0.00 |      0.16
---------------------+----------------------------------------------------------------------------------------+----------
               Total |        11        481        501        542        517        416         63          1 |     2,532
                     |    100.00     100.00     100.00     100.00     100.00     100.00     100.00     100.00 |    100.00
```

You may have noticed, however, that the cross-tabulation between `employst` and `gender` was also displayed. All the commands in the do file are run whenever you press the icon. If you would like to execute only one or a selected set of commands within the do file, you can highlight those commands using your mouse and then press the icon. To illustrate this process, highlight the command to produce the cross-tabulation between `employst` and `agecats` and press the icon. Notice that this time only the one cross-tabulation is displayed in the Results window.

A Closer Look: Do Versus Run

You may have noticed a very similar-looking icon directly next to the icon. This icon is what is called the "Run" button. Both buttons execute either the whole or selected parts of a do file. Pressing the icon, however, executes all the commands "silently," meaning nothing is displayed in the results window. The advantage of using the icon is that the commands are completed slightly quicker and the Results window is not filled with results you may not need to see. The icon is best to use when you have commands in a do file that you are sure accomplish exactly what you want and do not have any pertinent display (e.g., creating or recoding variables). Do not use the icon when your results are important, such as conducting statistical analyses.

Let's cover one more example to further emphasize both the utility and the method of using do files. In Chapter 2, you created a new variable `isol` that categorized young adults who had dated 2 or fewer people and had 2 or fewer friends as isolated. Now you have decided that you may need to change that definition to only classify young adults who have not dated and do not have any friends as isolated.

Begin by typing the commands to create the original variable into the do file. To create this variable, you had to use two commands: `gen isol=0` and `replace isol=1 if datnum<=2 & numfrien<=2`. When entering these commands into the do file, be sure that each goes on its own line by pressing **Enter** after you input each command line. Next, highlight both lines and press the icon. The new variable `isol` should be listed in the Variables window.

Next you need to create your altered, more restrictive indicator of being isolated. You might be tempted to alter the 2s in the command lines to 1s and rerun the commands. If you try this method, however, Stata will issue an error because the variable `isol` already exists in the data set, so you cannot create a new variable with the same name. Seeing this error message, you may either try dropping the existing `isol` variable or simply altering the new variable name in the command lines and then rerunning the commands. Although these quick fix options would work, they are not the most effective techniques.

A main advantage of do files is that they allow you to keep a history of all of your analyses. If you were to essentially overwrite the original commands that produced the `isol` variable, you would have no record that you ever tried defining isolated as having dated 2 or fewer people and having 2 or fewer friends. You may think that there is no way you will forget that fact, but it is not uncommon for research projects to take many months and even years. This detail, although prominent today, could be easily lost over the life span of the project. When you forget aspects of your project that you have conducted *and* have no record of them, you risk unnecessarily repeating commands and analyses.

For this reason, a more effective strategy for creating the new isolated variable is to highlight the two command lines and copy and paste them to two new lines in the do file. At this point, your do file should look like this:

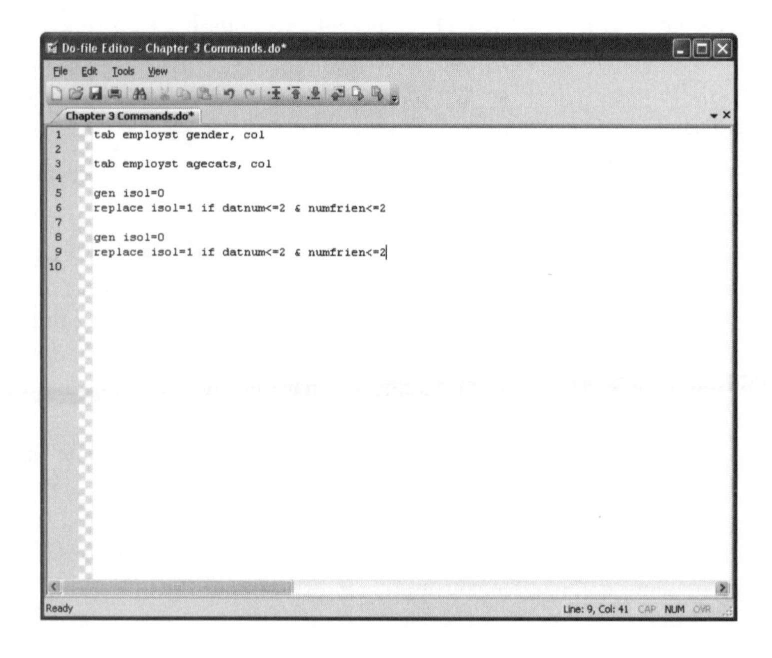

Notice, you can add blank lines to help keep your commands slightly more organized (more on this later).

In the second version of the `-gen-` and `-replace-` commands, you first need to alter *both* of the `isol` variable names. Because you have decided to only include people who have dated 1 or fewer people or have 1 or fewer friends, you might call this new variable `isol1`.[1] Once you have made these alterations, the new command lines in the do file should read: `gen isol1=0` and `replace isol1=1 if datnum<=2 & numfrien<=2`. At this point, you have changed the new variable name but not what it contains. If you ran these two commands right now, the new `isol1` variable would still be coded as 1 if the respondent had dated 2 or fewer people and had 2 or fewer friends. Consider what needs to be changed in the commands to achieve the desired result of the `isol1` variable being set to 1 if the respondent has dated 1 or fewer people and has 1 or fewer friends. You need to swap both of the 2s in the `-replace-` line with 1s. Now the command line is telling Stata to replace the `isol1` variable to equal 1 if both `datnum` and `numfrien` are equal to or less than 1. After you have made the necessary changes, your do file should look like this:

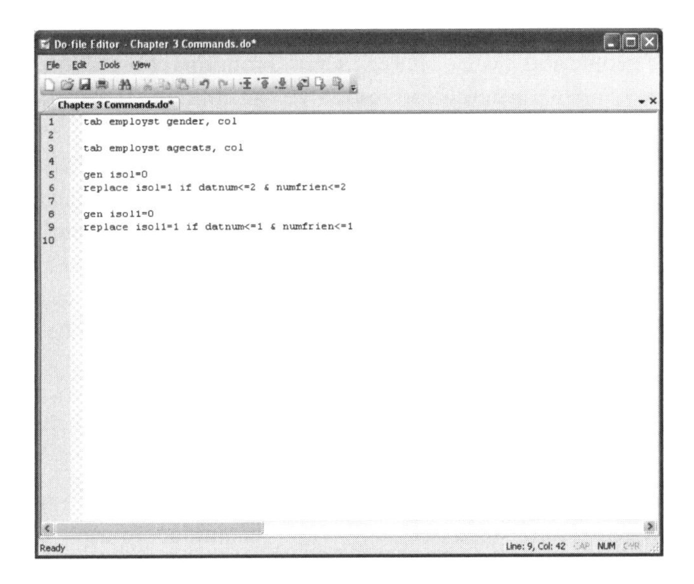

You only want to run the last two lines because you do not want to reproduce the first two cross-tabulations, and you already created the `isol` variable. Therefore, highlight the last two lines and press the icon. To see the difference between the two variables, type `tab isol isol1` into either the

[1]Here you see another good reason to use do files. It may be helpful to drop the original `isol` variable and change its name to `isol2` (to indicate 2 or fewer people dated and 2 or fewer friends). Then when you run the do file to create both variables, they are more easily distinguished.

Command window and press Enter (or into the do file and highlight and execute the line). The following results should be displayed:

```
. tab isol isol1

           |        isol1
      isol |         0          1 |     Total
-----------+----------------------+----------
         0 |     2,505          0 |     2,505
         1 |        26          1 |        27
-----------+----------------------+----------
     Total |     2,531          1 |     2,532
```

This cross-tabulation shows that the new variable `isol1` is much more restrictive than the original version. In fact, out of the 27 young adults who have dated 2 or fewer people and have 2 or fewer friends, only 1 has only dated 1 or fewer people and has 1 or fewer friends. From this comparison, you may decide that the new version is too restrictive and continue to use the original isolated indicator. This example shows why it is both helpful to use do files (retyping both commands to create the new variable would have been much more time-consuming and potentially more error prone) and to create multiple versions of a variable rather than dropping an old version and replacing. Had you dropped the original `isol` variable and changed the `-if-` statements in the do file to recreate the new version, you would have to repeat the process to go back to the original version, which turned out to be the more appropriate indicator.

GETTING THE MOST OUT OF DO FILES

Hopefully, at this point you are convinced that do files are a useful tool when conducting analyses in Stata and also feel confident in using their basic functions. This next section covers additional aspects of do files that will help you keep them and your analyses organized and effective.

Even with the relatively few commands you have entered into your current do file, you may notice that these files can get a bit messy pretty quickly. One way to alleviate this problem is by inserting notes into your do files. Notes can be used to create headings and to provide information about why commands are included. As mentioned above, Stata recognizes each new line as a separate command. To tell Stata that a particular line is a note and not a command, insert a "*" at the beginning of the line. For example, you could start your do file with a note that indicates the purpose of the do file. Create a new line at the top of the file by pressing Enter when your cursor is at the start of the very first line, then type "*Project Examining Employment/School Patterns." Similarly, you might create a heading that denotes that you are running cross-tabulations versus creating new variables. Or you can enter more detail about the creation of a particular variable. For example, you might type "*Generating new variable to indicate being isolated with 2 or fewer dated and 2 or fewer friends," on the line before the first `-gen-` command. Inserting notes in this way greatly aids your ability to keep track of what analyses and operations you have conducted.

Although you have not encountered such a situation yet, there will be times when you enter a command line that is very long. For example, you may produce summary statistics of numerous variables. These types of lengthy commands typically spill onto multiple lines when you enter them in the Command window. The Command window has no problem with these multiple lines because it treats everything that is inputted as a single command until the **Enter** key is pressed. A lengthy command in a do file, however, might be difficult to read if it extends further than the length of the Do File window. When you run into such a situation, you might be tempted to put the command line on two separate lines in the do file by pressing **Enter** somewhere in the middle of the command. But remember Stata sees a carriage return (pressing **Enter**) at the end of line that starts with a command as the end of that command. If you were to split a `-sum-` command into two lines by pressing **Enter** in the middle of the command, Stata would produce the statistics but only for the variables included on the first line.

To address this situation, you need to tell Stata that the information contained on the second line should be treated as part of the command on the preceding line. The symbol to indicate this carryover is three consecutive backslashes. When you come to the point of the command that you would like to break, insert `///` and then press **Enter**. If you highlight both lines and press the icon, Stata will treat both lines as one long command.

Finally, there are situations in which you want to keep the commands that you have used once but that you no longer need. In the example above, you decided not to use the `isol1` variable because it was too restrictive. It is helpful to keep the commands used to create this variable in the do file because you might forget that you had tried this categorization, leading to the type of unnecessary replication discussed above. At the same time, you might not want to continually create this variable that you are not using, both for clarity (i.e., it is easier to only have one variable that indicates being isolated in the data set) and file size concerns.

To balance both these concerns, you can leave a set of commands in a do file but then tell Stata not to execute them (even if they are highlighted or the entire file is run). To do so, you need to insert a backslash followed by an asterisk (`/*`) where you want to begin the omission and an asterisk followed by a backslash (*/) where you want the omission to end. When Stata sees the first symbol it essentially stops processing and running commands until it sees the second symbol. Although you can enter these symbols on the same line as the command(s) you do not wish to be executed, it is advisable to put them on their own separate lines, before and after the command(s). Putting them on their own lines makes them more prominent and will help you identify the areas of a do file that are not being run. So to "hide" the creation of the `isol1` variable, type /* on the line before the `-gen-` command line and a */ on the line after the `-replace-` command line.

Below is an example of an ideal do file for the commands that have been run thus far in the chapter, along with a lengthy multiple regression. This example illustrates how notes, lengthy command lines, and command omissions can be handled to produce a clear, informative, and well-organized do file.

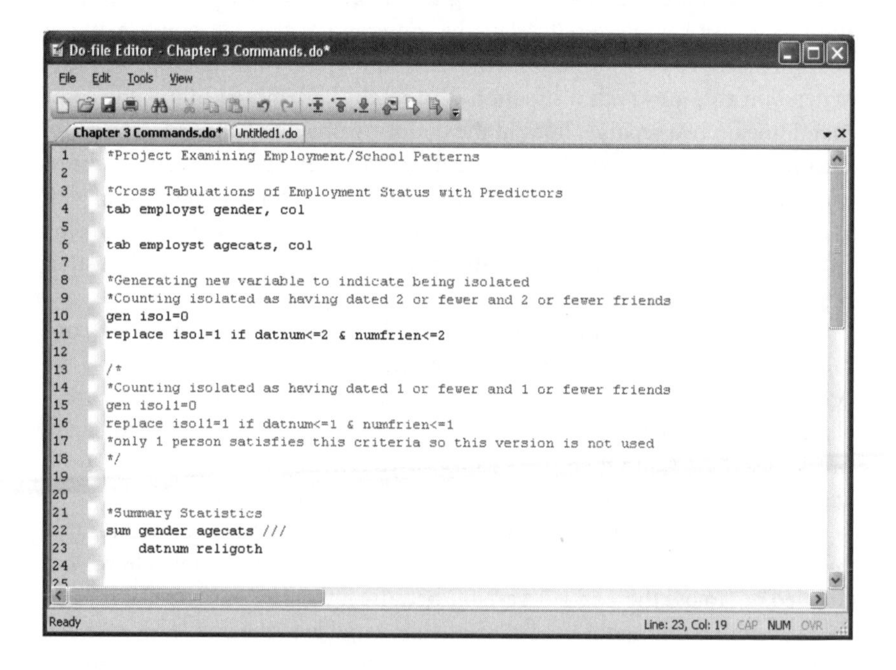

Notes have been inserted as both major headings and subheadings to provide information about the commands that follow. It can be especially helpful to include notes that provide further information about why and how a particular variable is created (e.g., the note explaining the criteria for classifying a person as isolated for the `isol` variable). Also, the creation of the `isol1` variable is omitted using the combination of the /* and */ before and after the commands. A note is included within this omission to explain why the variable is not used. Finally, the -sum- command command that was relatively lengthy has been split into two lines using the /// break to make it more readable.[2]

The rest of the chapter explains the execution of Commands through the Command window interface. Every command that is provided would be typed in exactly the same way into a do file. They can then be executed through the do file as shown above.

Data Management

One of the most significant aspects of quantitative research projects is data management. The term *data management* refers to all the operations that have to be completed to prepare the data so that they can be analyzed by statistical methods. One of the most common misperceptions among people who have never conducted quantitative analysis is that data sets come ready made for analyses. In reality, virtually every data set needs to undergo some alterations or adjustments for any particular research project. You already have conducted some central data management techniques, including the creation and alteration of variables. This section addresses several of the most important and frequently used operations. Some aspects of data management (such as assigning labels to values) are aimed at convenience, while others (primarily handling missing data) can have significant consequences for the validity of your analyses.

WORKING WITH LABELS

Chapter 1 introduced the idea of labels attached to variables and their values. It is helpful to know how to handle labels whether you are entering your own data or working with a secondary data set. Labels are used to provide more information about a variable and its values. When you begin working with larger data sets that contain numerous variables, the information contained in labels will prove invaluable.

There are two primary types of labels in Stata: variable labels and value labels. Variable labels are attached to the entire variable and are generally used to provide more detail about the variable itself. Value labels are labels that are attached to specific values within a variable. As was discussed in Chapter 1, most variables are stored as numbers, and those numbers refer to some category. For example, all respondents who report being "Male" are coded with the number 0 on the variable -gender-. In this case, "Male" is the value label

[2]In Stata 11 and 12, any portion of a do file that is not actually executed by Stata when the is pressed (i.e., a note or omitted command) is displayed in green.

attached to the number (i.e., value) 0. In Stata, a value label is the full set of individual labels attached to each value within a variable (e.g., Male and Female).

To demonstrate how to create and attach labels, consider the isol variable that you created earlier in this chapter. You may have already become slightly confused as to which classification criterion was used for which version of the isolated variable. Now imagine that you have worked on a research project for several weeks or months. Remembering exactly how this variable was created could be difficult. Using notes within do files, as shown above, is one strategy to address this problem, and attaching variable labels is another.

Think back to Chapter 2 and remember how you used intuition to derive the appropriate command name. If you consider how you might ask a smart colleague to label a variable, you should come up with the command to create and attach a label to a variable: -label variable-, shortened -lab var-. After you have entered the command into the Command window, you now need to tell Stata which variable you want to be labeled, followed by the label itself. For example, type lab var isol Isolated into the Command window and press **Enter**. Next produce a distribution of the isol variable (i.e., tab isol). The following results are displayed:

```
. tab isol

    Isolated |      Freq.     Percent        Cum.
-------------+-----------------------------------
           0 |      2,505       98.93       98.93
           1 |         27        1.07      100.00
-------------+-----------------------------------
       Total |      2,532      100.00
```

You can see that the variable label "Isolated" is displayed in the upper left-hand corner of the cross-tabulation. When you produce this distribution, this variable label will help remind you what you mean by the shortened phrase "isol."

The label "Isolated" is indeed helpful in this regard. But the creation of this particular variable was somewhat complex, and it would be helpful to have even more information in the variable label. Stata allows for very long (up to 80 characters) variable labels. If, however, the variable label contains spaces, you must enclose the label in quotations when typing it into the command. For example, you may want to list the criteria used to classify someone as isolated in the variable label. To do so, type lab var isol "Isolated with 2 few dates and friends" into the Command window and press **Enter**. If you use the -tab- command to create another distribution of the isol variable, you notice that this more detailed label is included in the upper left-hand corner of the table.

This example illustrates another important component of using the -lab var- command: New variable labels can be attached without first deleting an old variable label. That is, invoking the -lab var- command automatically overwrites any existing variable label.

Variable labels clearly help provide more information about the meaning of the variable. With the current label attached to the isol variable, however, it is still unclear whether respondents who are isolated are coded as being equal to 0 or to 1. Value labels help keep this type of distinction clear. The commands associated with value labels are similar to the ones you have used thus far. The main difference is that attaching value labels requires two separate commands—one to create the value label and one to tell Stata to which variables to attach that value label.

Keep in mind that by "value label," Stata is referring to the whole set of labels given to a set of values. For example, the combination of 0 referring to Male and 1 referring to Female would form the value label for the variable gender. The first step in attaching a value label is to define the value label. Defining the value label means you tell Stata what label should be attached to which numerals in that set. The command to define the label is -label define- (-lab def-). After you type the command name, you have to create a name for the value label set. For the present example, you could name the value label set isollab to indicate that it is the value label to indicate who is isolated. To this point, the command would read lab def isollab.

Now you must indicate what values should be associated with what label. In the current example, cases that are coded as 0 are defined as not being isolated, whereas cases coded as 1 are the ones that are defined as isolated. The command to complete the definition is similar, in form, to the variable label command, except that you are giving each value a label. Type the command: lab def isollab 0 "Not Isolated" 1 "Isolated" in the Command window and press **Enter**. Notice that the value is typed first, followed by the label. Also, as with variable labels, if there is a space in the label, it must be enclosed in quotes. Even though the label associated with 1 (i.e., "Isolated") does not have a space in it, it may be easier to simply always enclose labels in quotes so that you do not forget to do so when there are spaces in the label.

At this point, you have only defined the value label but have not attached that value label to any variable. If you produce a distribution of the isol variable, you see that 0 and 1 are still displayed instead of the labels. The command to attach the value label is similar to the previous two: -label value- (-lab val-). This command tells that Stata is going to be attaching a value label. After the command, you then need to tell Stata which variable is going to receive the value label followed by which value label to use. In other words, you are telling Stata (or your smart colleague) to "label (the) values (of) variable name (with the value label) value label name." Type lab val isol isollab into the Command window and press **Enter**. Next produce a distribution of the isol variable (i.e., tab isol). The following results show that the value label was attached correctly. The 27 cases that were classified as Isolated are now labeled as such.

```
. tab isol

    Isolated |      Freq.     Percent        Cum.
-------------+-----------------------------------
Not Isolated |      2,505       98.93       98.93
    Isolated |         27        1.07      100.00
-------------+-----------------------------------
       Total |      2,532      100.00
```

At first, it may seem that this two-part command to attach value labels is inefficient. You might wonder why it is not possible to define and attach the value label all in one step. This situation is an instance where what in a specific case seems cumbersome is actually quite efficient in the context of a full research project. In the example, you defined a value label for a variable that contains cases that are categorized as not being isolated and cases that are classified as isolated. If this were an actual research project, the value label isollab could be used to label the values of any variable that contains these two groups, as long as those two groups are coded 0 and 1, respectively. For example, you could label the variable you created earlier, isol1, with the same value label. Typing lab val isol1 isollab into the Command window and pressing **Enter**, produces the following results:

```
. tab isol1
       isol1 |      Freq.     Percent        Cum.
-------------+-----------------------------------
Not Isolated |      2,531       99.96       99.96
    Isolated |          1        0.04      100.00
-------------+-----------------------------------
       Total |      2,532      100.00
```

Notice that even though the specific cases are coded differently, the label attached to the values has been correctly applied. If defining and attaching the value label was contained in one command, the entire value label would have to be typed every time you wanted to use it. Certain value labels, such as Yes–No or Agree–Disagree, might apply to several variables in a data set. Once you have defined the value label for one of these common sets, you can attach it to as many variables as it applies.

The label attached to a particular value only pertains to that value label. That is, in the value label isollab, you have defined 0 to be associated with the label "Not Isolated." Any variable that you attach the isollab to will have the label "Not Isolated" applied to any case that is coded 0. But you can create a different value label in which the value of 0 has a different label. For example, you could create the value label yesno which is defined such that 0 is associated with "No" and 1 with "Yes." You can then use this value label for any variable in which cases coded as 0 should be labeled as "No."

Additionally, value labels are simply that: labels. The numeric values are the real codes that Stata uses when conducting analyses. You could apply several different value labels to the same variable (each new one would automatically overwrite the existing value label), and the numeric values would never change. This feature is beneficial because you can alter the value labels as often as you like without having to actually change the way in which the respondents are coded. For example, you may decide that "Isolated" is too strong of a term and prefer to label these cases as "Estranged." You could create a new value label (lab def estrange 0 "Not Estranged" 1 "Estranged") and then attach it to the isol variable (lab val isol estrange). If you produce the distribution of this variable, 27 cases are still grouped together and are still coded as 1, but now they are simply labeled as "Estranged" rather than "Isolated."

There are situations, however, in which you need to know the exact value, not the label, of a particular group. For instance, you might want to create a distribution of employment status of young adults if the respondent is not counted as being isolated. As you learned in Chapter 2, you need to use an -if- statement to produce this distribution correctly. When you go to type the command you would enter tab employst if isol== #. With the value labels attached to the isol variable, it is difficult to know what number should replace the "#." The easiest way to identify the correct numeral is by producing two distributions of the isol variable, one that displays the value label and one that displays the numeric values associated with each label. First, type tab isol into the Command window and press **Enter**. By default, the -tab- command produces a distribution showing the value labels, as presented just above. To produce a distribution that shows the underlying numeric values, you need to invoke the -nolabel- (-nol-) option. Type tab isol, nol into the Command window and press **Enter**. The following results are displayed:

```
. tab isol, nol

   Isolated |      Freq.     Percent        Cum.
------------+-----------------------------------
          0 |      2,505       98.93       98.93
          1 |         27        1.07      100.00
------------+-----------------------------------
      Total |      2,532      100.00
```

You can see that the distribution is exactly the same, only the labeling of the values has changed. Now you can clearly see that people who are not classified as isolated are coded as 0. Therefore, to produce the desired distribution of employment status for cases who are not isolated, you would type `tab employst if isol==0` into the Command window and press **Enter**.

MISSING DATA

Whenever surveys are conducted, it is almost inevitable that some cases will be missing, meaning they do not provide a response, for particular questions. Frequently missing data are caused by people either not knowing or not wanting to provide an answer for a question. In other situations, missing data may have been designed into the survey, such as not asking a respondent how many of his or her friends are religious if he or she previously reported not having any friends. Whatever the cause of the missing responses is, it is vital that these responses be handled in a deliberate manner; otherwise, the statistical analyses may produce unwanted results.

Consider that you were trying to predict the number of friends that young adults report to have. You might hypothesize that as young adults get older their number of friends will increase. To test your prediction, you decide to estimate a linear regression with `numfrien` as the dependent variable and `agecats` as the independent variable. Before you execute this regression, you think it is a good idea to see a distribution of how many friends the respondents report. Typing `tab numfriend` in the Command window and pressing **Enter** produces the following table:

```
. tab numfrien

 (numfriend_w3)  |
   N:1. Now for  |
the next set of  |
questions, I'll  |
  be asking some |
          things |      Freq.     Percent        Cum.
-----------------+-----------------------------------
               0 |          6        0.24        0.24
               1 |         52        2.05        2.29
               2 |        165        6.52        8.81
               3 |        517       20.42       29.23
               4 |        515       20.34       49.57
               5 |      1,265       49.96       99.53
         Refused |          3        0.12       99.64
 Legitimate skip |          9        0.36      100.00
-----------------+-----------------------------------
           Total |      2,532      100.00
```

You see that there are two categories at the bottom of the table that refer to missing data: `Refused` and `Legitimate Skip`. The `Refused` group contains respondents who did not want to report how many friends they have, while the `Legitimate Skip` group are respondents who were not asked this question based on the design of the survey. Because they did not provide an answer, it is impossible to know how many friends these 12 cases have. There are advanced statistical techniques for handling this type of missing data, but perhaps the most common method is to simply eliminate these cases from any analyses.

Stata contains a set of "missing codes" that tells it that the cases have not provided an answer to a particular variable. When cases are set to one of these missing codes, Stata automatically eliminates them from any statistical analyses performed using the variables entered into the command. Therefore, you need to replace the 12 cases in the `numfrien` variable with one of these missing codes.

As covered in Chapter 2, there are a few different methods that would accomplish this goal, but for the purpose of identifying missing cases, the `-recode-` command will be most effective. You might consider the `-replace-` command. As you can see, there are multiple values to be changed, which would require multiple `-replace-` commands but only one `-recode-` command. Furthermore, this is one instance where the `-gen(newvar)-` option probably does not need to be invoked. As will be shown, each missing code has a specific value for each reason a case may be missing, allowing the user to identify and even revert back to the original values if needed.

To use the `-recode-` process, you first must know what the current values of the missing cases are. This is another instance in which the `-tab-` command with the `-nol-` option is useful. Type `tab numfrien, nol` into the Command window and press **Enter**.

```
. tab numfrien, nol

   (numfriend_ |
     w3) N:1.  |
  Now for the  |
  next set of  |
   questions,  |
      I'll be  |
  asking some  |
       things  |      Freq.     Percent        Cum.
---------------+-----------------------------------
            0  |          6        0.24        0.24
            1  |         52        2.05        2.29
            2  |        165        6.52        8.81
```

(Continued)

(Continued)

```
          3 |        517       20.42       29.23
          4 |        515       20.34       49.57
          5 |      1,265       49.96       99.53
        888 |          3        0.12       99.64
        999 |          9        0.36      100.00
------------+-----------------------------------
      Total |      2,532      100.00
```

This distribution shows that cases who refused to answer the question are coded as `888`, and cases that were legitimately skipped out of the question are coded as `999`. Had you run the regression analyses on this version of the `numfrien` variable, Stata would have treated 3 cases as having `888` friends and 9 cases as having `999` friends, which clearly would cause some problems for the regression estimation and its subsequent results.

Stata has 27 different missing codes that can be used to define a value as missing for a particular reason. The most basic code is a ".", which is sometimes referred to as system missing. System missing means that the case is just missing, with no particular reason known or identified. The other 26 codes are similar but the "." is followed by a letter of the alphabet, such as .d, .s, or .r. The letters can be used as shorthand for why the case is missing on the variable. Remember that Stata will treat all cases that are given any of the 27 missing codes exactly the same. The benefit of using specific codes, therefore, is so that the user can identify, and potentially use, the specific reason why a case is missing.

A Closer Look: When a Missing Case May Not Be Missing

It may seem easier to simply set all missing cases as the default "." code, as opposed to distinguishing each type of missing case with a unique code. In certain situations using the generic system missing code may work, but generally if the specific reason why the case is missing is known, it is best to use a more informative missing code.

For example, in the example with the `numfrien` variable, it may not seem important whether a case is missing because the respondent refused to answer or was skipped out of the question. For a particular analysis, however, it might be advisable to create a new variable in which all the `Legitimate Skip` cases are coded as having 0 friends. This decision would be particularly defensible if the cases were skipped because they said they never talk to anyone else on a previous question. Similarly, if there were cases that were missing on this variable because they responded that they "Don't Know" how many friends they have, it may be reasonable to argue that these cases should be coded as having 0 friends (i.e., not knowing is similar to not having friends). If all the missing cases were given a similar missing code, it would be impossible to make these distinctions. Even more important, if at a later point these decisions on how to treat particular missing cases are reversed or changed, using specific missing codes allows for each type of missing case to be identified and recoded as needed.

For the `numfrien` variable, you might decide to set the cases who refused to answer the question to `.r` and those that were legitimately skipped to `.s`. To execute the recode, type `recode numfrien (888=.s) (999=.r)` in the Command window and press **Enter**. Now if you type `tab numfrien` in the Command window and press **Enter**, the results should look as shown below:

```
. tab numfrien

 (numfriend_w3) |
   N:1. Now for |
the next set of |
questions, I'll |
  be asking some |
          things |      Freq.     Percent        Cum.
----------------+-----------------------------------
              0 |          6        0.24        0.24
              1 |         52        2.06        2.30
              2 |        165        6.55        8.85
              3 |        517       20.52       29.37
              4 |        515       20.44       49.80
              5 |      1,265       50.20      100.00
----------------+-----------------------------------
          Total |      2,520      100.00
```

Stata by default does not include cases that have been assigned a missing code in any analyses, which is why the 12 cases that were recoded are not shown in the distribution of the `numfrien` variable.

It is important to see that the percentages in the distribution have changed because the total number of cases used to calculate those percentages has changed from 2,532 to 2,520, due to the 12 missing cases being excluded. There might be a reason that you want the percentages to be calculated using the entire sample, not only respondents with valid responses on the variable. To make Stata show the cases that are missing on a variable in the distribution (and adjust the percentages accordingly), the -missing- (-mis-) option can be invoked. Type tab numfrien, mis into the Command window and press **Enter**.

```
 (numfriend_w3) |
   N:1. Now for |
the next set of |
questions, I'll |
 be asking some |
         things |      Freq.     Percent        Cum.
----------------+-----------------------------------
              0 |          6        0.24        0.24
              1 |         52        2.05        2.29
              2 |        165        6.52        8.81
              3 |        517       20.42       29.23
              4 |        515       20.34       49.57
              5 |      1,265       49.96       99.53
             .r |          9        0.36       99.88
             .s |          3        0.12      100.00
----------------+-----------------------------------
          Total |      2,532      100.00
```

This new distribution is identical to the first one you produced, but the codes of the 12 missing cases have been changed. You can see, however, that if you were interested in reporting the percentage of young adults who report having 5 friends, the figure would be slightly different (50.20 vs. 49.96) based on whether you include the missing cases or not.

Although Stata does not use cases that have been assigned missing codes in analyses, it does not completely exclude these cases from all commands. Most important, when using -if- statements, cases that are missing are still evaluated. Specifically, Stata treats all missing values as being of a value greater than the highest nonmissing value on that variable. This fact is important to keep in mind whenever a variable that has missing values is referenced in an expression containing a greater than (>) symbol.

An example should clarify this point. Imagine that you want to create a dichotomous variable that indicates young adults with many friends and those with few friends. To make this classification, you decide that anyone with 4 or more friends should be considered as having many friends. The -recode- command along with its -gen(newvar)- option would work to create this new variable, but for the purposes of the example the combination of the -gen- and -replace- command is more illustrative.

First, to create the new variable, type gen hifrien=. in the Command window and press **Enter**. This method of creating a new variable in which every case is first set as missing is an effective way to prevent including unwanted cases (e.g., missing) in a newly generated variable. Next, set all the people with 3 or fewer friends as 0 by typing `replace hifrien = 0 if numfrien<=3` in the Command window and pressing **Enter**. Now all the people with 4 or more friends need to be coded as 1 on the new variable. It might seem to make sense to use a similar command as above with a greater than sign. That is, to type `replace hifrien = 1 if numfrien>=4`. If this command was used, the new variable's distribution would be

```
. tab hifrien
    hifrien |      Freq.     Percent        Cum.
------------+-----------------------------------
          0 |        740       29.23       29.23
          1 |      1,792       70.77      100.00
------------+-----------------------------------
      Total |      2,532      100.00
```

This new variable would not be correct because the distribution of `numfrien` showed that 12 cases were missing so that the total cases should only be 2,520. The reason that the missing cases were incorrectly coded using the previous command was that Stata interprets the clause "`if numfrien>=4`" to mean any case that has a value of 4, 5, *or* missing. The missing is included because Stata treats all missing values as being equal to a value greater than the highest nonmissing value (here 5). To correct this mistake, the command should be typed as `replace hifrien=1 if numfrien>=4 & numfrien<=5`. Once this command has been executed, the distribution of the new variable will look like the following:

```
. tab hifrien
    hifrien |      Freq.     Percent        Cum.
------------+-----------------------------------
          0 |        740       29.37       29.37
          1 |      1,780       70.63      100.00
------------+-----------------------------------
      Total |      2,520      100.00
```

In this version, the missing cases have remained missing. For this particular replacement, the combination of both the greater than and less than expressions was not completely necessary, as it would have been possible to use an equal to 4 or equal to 5 `-if-` statement as well. In general, when using an `-if-`, or a similar expression, that uses a greater than symbol, it is advisable to also include a less than (the maximum value) statement to ensure that the missing cases are not included unnecessarily.

The analytic strategies used to handle missing data are too complex for the purpose of this book. What is vital to understand is how Stata handles missing data by default. If cases that should be missing are not coded with missing codes, then they are included in all analyses. When they are set to a missing code, then they are automatically excluded from all analyses. If you are entering data by hand, you can use missing codes as you enter the data rather than recoding them later. Some data sets already have missing cases set to a missing code. The diversity with how missing data are handled when creating data sets, however, is why it is a good data management practice to always look at a distribution table of the variables you are including in an analysis.

USING STRING VARIABLES

Chapter 1 introduced numeric and string variables and, as was noted, Stata treats them very similarly in most instances. There are, however, a few key commands that are slightly different when string variables are used.

As shown in Chapter 2, the `-gen-` command by default creates a numeric variable. To force Stata to generate a string variable, you need to specify that you want to create such a variable. For example, you might want to generate a variable, named `christoth`, that indicates whether respondents identify the generic "Christian" as their other religion. The first part of the command is the same: `gen`. Next, however, you have to indicate the variable you are creating should be a string. Again, thinking about how you would say this command might lead you to consider "generate (the) string (variable) christoth." This intuitive thinking translates very closely to the correct command: `gen str christoth`. The only addition is the `-str-` to designate that the `christoth` variable should be stored as a string variable.

The second difference follows the equal sign. Whenever you refer to a value in a string variable, it must be surrounded by quotation marks. Therefore, if you want the new variable to be coded "Christian," you would type `gen str christoth="Christian"`. If you press **Enter** at this point, every case will be coded as Christian, meaning you need to use an `-if-` statement with the `religoth` variable.

Because the `religoth` variable is a string variable, you need to use the quotation rule just discussed. When you refer to the value of a string variable, it must be surrounded by quotation marks. The full command to create the desired variable is: `gen str christoth="Christian" if religoth== "CHRISTIAN" | religoth== "JUST CHRISTIAN" | religoth== "NONDENOMINATIONAL CHRISTIAN"`. This command generates a variable that codes cases to equal "Christian" if the case responded with one of the general Christian categories on the `religoth` variable.

The spelling and capitalization of string values must be exact. For instance, if you accidently typed "JST" instead of "JUST," the 19 cases who reported being "Just Christian" on the `religoth` variable would not have been counted as fulfilling the `-if-` clause in this `-gen-` command. The capitalization of string variables does not matter in any substantive sense (i.e., Stata does not care what the capitalization is), but whatever the capitalization is used in setting string variable values must be consistently followed.

The command to display the distribution of this string variable is exactly the same as if it were a numeric variable. Typing `tab christoth` in the Command window and pressing **Enter** produces a table similar to those produced with numeric variables:

```
. tab christoth

 christoth |      Freq.     Percent        Cum.
-----------+-----------------------------------
 Christian |         44      100.00      100.00
-----------+-----------------------------------
     Total |         44      100.00
```

A total of 44 cases reported one of the three general Christian categories in the `religoth` variable and have been set to "Christian" on the new string variable you generated. All the other cases have been set to missing.

The way in which these missing data are handled is one other aspect that is unique to string variables. As noted above, there are 27 missing codes that can be used to identify cases with missing data on numeric variables. These codes can only be used with numeric variables. Because the values of the string variables are the actual text, Stata will interpret a missing code (e.g., .d or .s) on a string variable as a valid value. The only valid missing code on string variables is a blank, that is "".

As a final illustration of using string variables, the `christoth` variable needs to be completed. To do so, you need to replace all the currently missing cases who provided a particular denomination to equal Specific Other Denomination. `-replace-` is the most effective command in this situation because `-recode-` is one command that does not work with string variables. Just as with the `-gen-` command, the basics of `-replace-` are similar for numeric and string variables. The main difference is that when referencing values for string variables, you must again enclose them in quotes. To change the missing value type `replace christoth="Specific Other Denomination" if christoth=="" & religoth!=""` in the Command window and press **Enter.** Notice that the quotes surround the values of the string variable, not the variable name. This command is telling Stata to code all cases who are currently missing on christoth but *not* missing on religoth (i.e., they reported a specific "other" denomination) to "Specific Other Denomination." Note that if the second clause in the -if- statement was not included, cases who were missing on religoth would also be set as "Specific Other Denomination." This coding would be inappropriate because these cases were missing because they did not report an "other" denomination.

This -replace- command illustrates two other noteworthy points. First, when referencing missing values in string variables, spaces matter. By default, missing cases on string variables are an actual blank or nothing. Stata would see a space between the quotes (i.e., " ") as an actual character and would not have properly replaced the values. Second, you can reference the variable that is having values replaced in the -if- statement. You could have listed every single other possible denomination from the religoth variable, but it is much quicker to simply specify the values of the new christoth variable (i.e., those who had not been set to "Christian") that needed to be replaced. This self-referencing with -if- statements works with all commands and variable types when appropriate.

Producing a distribution of the variable shows that the replacement was correct (tab christoth):

```
. tab christoth

                   christoth |      Freq.     Percent        Cum.
-----------------------------+-----------------------------------
                   Christian |         44       16.67       16.67
Specific Other Denomination |        220       83.33      100.00
-----------------------------+-----------------------------------
                       Total |        264      100.00
```

The previously missing cases have been set to the appropriate string value. The -tab- command displays values of string variables in alphabetical order by default, but the -sort- option can be invoked to have it list the values in frequency order. Finally, a string value can contain up to 244 characters, although the -tab- command will not show all the characters for extremely lengthy string values.

The determination of using a string versus a numeric variable should be based on what values are to be held by the variable. The primary advantage of string variables is that the exact response can be entered as the value. String variables, therefore, are a good option when the variable comes from an open-ended question.

A good rule of thumb, however, is to only use string variables when they are absolutely necessary. Even if a variable contains values that seem like text (e.g., Male), these values usually can be coded as numbers. In the example shown above, the christoth variable would be more effective as a numeric variable that uses value labels to identify the categories. Although many commands operate similarly from string and numeric variables, a major portion of statistical analyses require categories that are coded as numbers, making numeric variables generally the more effective option.

If you are using a secondary data set, the choice of using a string versus a numeric variable may not be up to you. It is not completely uncommon for variables in certain data sets to be stored as string variables even if they would be better served as numeric variables. Fortunately, Stata has a command that automatically changes string variables into numeric variables. Unfortunately, this is one instance in which intuition may not be terribly helpful.

The -encode- command takes a string variable, turns the categories into numerical values, and applies the string values as a value label. When using -encode-, the -gen(newvar)- option must be invoked to tell Stata what the name of the new numeric variable should be. If you wanted to change the string variable christoth to a numeric variable, you would type encode christoth, gen(nchristoth) in the Command window and press **Enter**. The variable name typed in the -gen(newvar)- option must be different from the string variable it is replacing. In the example, an "n" has been added to the beginning of the variable name to denote that this new variable is a numeric version of the previous string variable. If you were to produce a regular distribution of the new, numeric version of the variable, it would look identical to the table using the string version above. Invoking the -nol- option with the -tab- command reveals that the previous string values are now simply value labels. Type tab nchristoth, nol into the Command window and press **Enter**:

```
. tab nchristoth, nol

 nchristoth |      Freq.     Percent        Cum.
------------+-----------------------------------
          1 |         44       16.67       16.67
          2 |        220       83.33      100.00
------------+-----------------------------------
      Total |        264      100.00
```

The table shows that the distribution of cases is exactly the same as it was in the string version of the variable. Notice that the -encode- command assigns numeric values (starting at 1) in alphabetical order based on the string variable.

SAVING RESULTS

In addition to saving the commands that you use throughout your data analyses, it is clearly important to save the results that you produce. As you have probably noticed, every time you close Stata, the Results window is erased, destroying any tables or statistical analyses that you have produced. There are two primary methods for saving results in Stata.

The first of these methods is simply copying and pasting the results from the Results window into another software program. Any of the results, including tables and statistical analyses, can be copied from the Results window and pasted into either a word processing or database program. To copy the results into a word processing program, simply highlight the desired results with your mouse and click Ctrl+C. You can then paste these results into the program, using either Ctrl+V or right clicking on your mouse and selecting Paste. Copying tables into a database or spreadsheet program is similar, except once you have the table highlighted, right click on your mouse and select Copy as Table. Then select a cell in your database or spreadsheet and press either Ctrl+V or right click on your mouse and select Paste. The results may not transfer perfectly and might require some manipulation in the word processing or database software. Despite this disadvantage, this copy and paste method is a very quick and easy way to save results from Stata.

The second method involves using what are called "log files." These files keep a running log of anything that appears in the Results window after the log file has been started. Stata does not start a log file by default. Rather you must decide to start a log file and subsequently close it. The easiest way to start logging your work is to click on the icon. Log files can be saved either as a Formatted Log (.smcl) or Log (.log) format. The Formatted Log has its advantages if you are familiar with the .smcl format, but if you are not, the Log format may be more effective, as it is easily opened and readable by most word processing programs. To save a log in the Log format, choose "Log (.log)" from the drop-down menu under "Save as Type." After you have named your log file and clicked Save, a note in the Results window shows that a log file has been started. Everything that appears in the Results window after this point and before the log is closed is saved by the log file. If you have produced results that are still in the Results window but appear before you started the log, they will *not* be included in the log file. This facet is important to keep in mind because log files are not equivalent to saving the Results window. Instead they are more like a tape recorder that only captures what occurs between when "REC" and subsequently "STOP" are pressed.

To close the log file, press the icon again. When you do, a dialog box with three options, like the one shown below, appears. The first option "View Snapshot of the Log File" shows you what the log file has captured to that point. The second option "Suspend Log File" allows you to pause what is being recorded, conduct some "off the record" work, and then start recording again at a later point. The final option "Close Log File" stops the log file. Anything that you do after this point is not included in that log file. After you have checked the radio button next to the desired option, press OK.

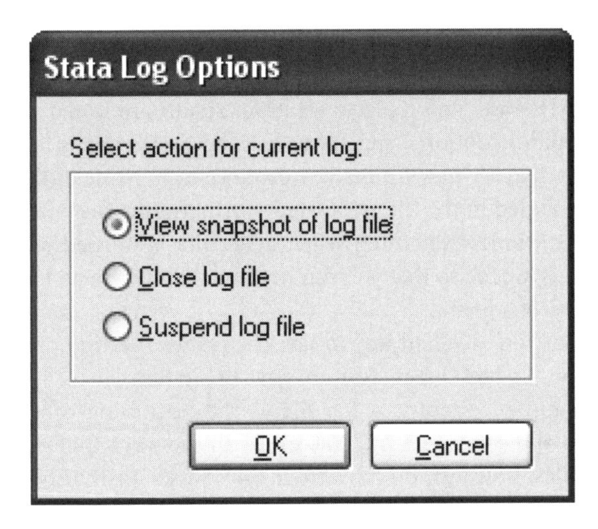

A nice feature of log files is that they can be reopened and appended. Using this option allows you to keep a continuous record over multiple Stata sessions. To reopen a log file, press the icon and then locate an existing log file. You might need to change the file type in the "Save as Type" drop-down menu to find the correct file. Once you have located the correct file, highlight the file and either double click or press Save. Another dialog box with three options, as shown below, appears:

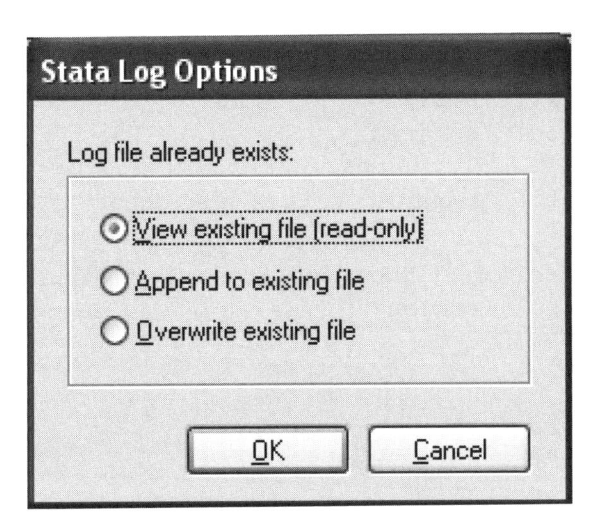

The first option, "View existing file (read-only)," displays all the results that are currently saved in the log file. The final option, "Overwrite existing file," completely erases and replaces whatever results are stored in the selected log file. The middle option, "Append to existing file," reopens the existing log file and adds whatever new results you produce to it. None of the results that are already included in the file are altered. Furthermore, the notes that appear in the Results window denoting when a log file is opened and closed are included in this log file so that you can keep track of where and when the new set of results were added.

Log files are an excellent way to save the results that you produce during a Stata session. In fact, when you are first using Stata it may be helpful to establish a habit of opening a log file at the beginning of every session. Doing so will allow you to have a record of all the work that you have completed. Log files, however, do have their drawbacks. Primarily, logs capture everything that appears in the Results window, including potential mistakes or unwanted results. After several sessions, log files can become quite lengthy, and it may be difficult to find the exact results you need. For this reason, it is helpful to use do files in addition to log files. Do files keep track of all the commands that you have used, and the insertion of notes can make it very easy to find the parts that you decide to use in your final project. Once you have determined which analyses you need, you can then open a log file and selectively run these commands through the do file. This combination of do files and log files is the most effective and efficient way to save your work in Stata.

Summary of Commands Used in This Chapter

```
*Working with Labels
lab var isol Isolated
tab isol
lab var isol "Isolated with 2 few dates and friends"

lab def isollab 0 "Not Isolated" 1 "Isolated"
lab val isol isollab

tab isol

lab val isol1 isollab
tab isol1
```

```
tab isol, nol

*Missing Data
tab numfrien
tab numfrien, nol
recode numfrien (888=.s) (999=.r)
tab numfrien
tab numfrien, mis

gen hifrien=.
replace hifrien=0 if numfrien<=3
replace hifrien=1 if numfrien>=4 & numfrien<=5


*String Variables
gen str christoth="Christian" if religoth==
"CHRISTIAN" | religoth== "JUST CHRISTIAN" | religoth==
"NONDENOMINATIONAL CHRISTIAN"

tab christoth

replace christoth= "Specific Other Denomination" if
christoth== "" & religoth!= ""

tab christoth

encode christoth, g(nchristoth)

tab nchristoth, nol
```

Exercises

Use the original `Chapter 3 Data.dta` for the following problems.

1. Open a log file and name it "Chapter 3 Exercise Results" to save the results of these exercises.
2. Open a do file and save it with the title "Chapter 3 Exercises". Complete the remaining exercises by typing them into this do file and then executing them.
3. Add a note to the top of the do file that indicates the purpose of the do file.

4. Create a new variable that categorizes the number of people dated (`datnum`) into three categories. Define the categories as people who have dated 2 or fewer people, between 3 and 10 people, and between 11 and 100 people. Name the new variable, `datlevsalt`.

5. Assign the variable label "Categories of People Dated" to the newly created `datlevsalt` variable.

6. Assign the following labels, Minimal Dating, Moderate Dating, and Extensive Dating, to the appropriate values of the `datlevsalt` variable.

7. Change the cases coded as "Legitimate Skip" on the `employst` variable to a missing code.

8. Produce a distribution table that does not show the missing cases on the `employst` variable and one that does.

9. Generate a string variable, named `mormon`, that distinguishes cases responding some form of Latter-Day Saints or Mormon on the `religoth` variable versus those who do not.

10. Convert the string version of the `mormon` variable into a numeric variable.

11. Close and save the log file.

PART II

Quantitative Analysis With Stata

4

Descriptive Statistics

At this point, you should feel fairly comfortable and confident with the basic operations of Stata. The next four chapters explain how to conduct basic quantitative analyses using Stata. The strategies and techniques covered are most commonly found in introductory social statistic textbooks. Stata can perform countless additional and more advanced statistical techniques than can be covered. The commands that are discussed, however, provide a foundation for understanding virtually any specific analyses one could perform in Stata. Then the Stata Help Files section of Chapter 8 addresses ways in which interested readers can help themselves learn how to conduct techniques that are not specifically covered.

To be clear, this book does not seek to explain how or why particular statistical analyses should be used. It should not be viewed as a replacement to a thorough quantitative analysis text. Rather it has been designed to act as a companion to such texts. Once you have a solid understanding of the mechanics underlying the basic analytic strategies, these chapters guide you through how to conduct them on real data. Most importantly, the interpretation of the results is extremely brief and is aimed at identifying the most important components of the output rather than explaining their meaning. Again, readers should consult dedicated quantitative analysis texts for a more comprehensive explanation of how to interpret the statistical figures presented.

Perhaps the most important first step in any quantitative research project is to understand the distribution of each of the pertinent variables. This chapter covers the commands that are used to produce univariate (i.e., single-variable) descriptive statistics, including measures of central tendency and variability. Methods for presenting these measures graphically are also addressed. All the examples that follow use the `Chapter 4 Data.dta`, available at **www.sagepub.com/longest**. This data set includes the full National Study of Youth and Religion (NSYR) Wave 3 sample of 2,532 young adults. All the missing cases have been replaced with appropriate missing codes in this data set, with .d referring to a response of "Don't Know," .r to a response of "Refused," and .s to cases that were legitimately skipped out of a question based on the survey design (i.e., skip pattern).

Frequency Distributions

For variables with a limited number of categories, the easiest way to gain an initial sense of how the variable is distributed is by producing a frequency distribution. A frequency distribution shows how many cases and the percentage of cases that belong to each category of a variable.

For example, you might conduct a project on young adults' perception and satisfaction with their physical appearance. The NSYR has a question that asked the respondents, "In general, how happy or unhappy are you with your body and physical appearance?" The responses to this question provide a good overview of young adults' body image. As was introduced in Chapter 2, the -tab- command is used to produce a frequency distribution of a variable. The variable for the question about body image is called body. Typing tab body into the Command window and pressing **Enter** produce the following results:

```
   (body_w3) P:3. |
  In general, how |
happy or unhappy |
     are you with |
    your body and |
           physic |      Freq.     Percent        Cum.
-----------------+-----------------------------------
     Very unhappy |         68        2.70        2.70
Somewhat unhappy |        389       15.42       18.11
          Neither |        234        9.27       27.39
   Somewhat happy |        953       37.77       65.16
       Very happy |        879       34.84      100.00
-----------------+-----------------------------------
            Total |      2,523      100.00
```

By now you have seen several such distributions, but it may be helpful to review the primary components in terms of how they would relate to an actual analysis. First, the variable label is listed in the upper left-hand corner of the table. For the body variable (and with all variables in the NSYR data set), the default variable label is a brief description of the survey question wording. Next, the left-hand column lists all the categories for the variable to which at least one respondent belongs. There might be other possible answer choices, but they are not shown in the frequency table if no one in the data belongs to those categories.

Finally, the three columns to the right display the number of respondents in each category (`Freq.`), the percentage of respondents in each category (`Percent`), and the cumulative percentage of respondents in each category (`Cum.`). The frequency is a count of all the participants who answered in each category. For example, 68 of the respondents reported being "very unhappy" with their physical appearance. The percentage reports this frequency divided by the "Total" number of valid (i.e., nonmissing) cases on this variable. Only 2.7% cases, of the 2,523, report being very unhappy with their body. This figure suggests that only a small set of young adults are very dissatisfied with their physical appearance. The cumulative percentage displays the total percentage of cases that belong to that category and the ones preceding it. For the `body` variable, 18.11% of the cases report being somewhat unhappy or very unhappy with their physical appearance. Again, you can see that the vast majority of young adults are either indifferent or at least somewhat happy with their physical appearance.

As was discussed in detail in the Data Management: Missing Data section of Chapter 3, the `-tab-` command by default does not list cases that did not provide an answer to that variable's question (i.e., are missing). Furthermore, the percentages that are displayed in the table are based on the total number of cases who provide a valid answer to the particular variable, not the total number of cases in the data set. If you are interested in the percentages of respondents in each category based on the total sample of 2,532, you can add the `-mis-` option to the previous `-tab-` command. Type `tab body, mis` in the Command window (or press **Page Up** while the Command window is highlighted and add "`, mis`" to the previous executed command) and press **Enter**. The full sample results appear as follows:

```
   (body_w3) P:3. |
  In general, how |
happy or unhappy |
     are you with |
    your body and |
           physic |      Freq.     Percent        Cum.
------------------+-----------------------------------
     Very unhappy |         68        2.69        2.69
Somewhat unhappy |        389       15.36       18.05
          Neither |        234        9.24       27.29
   Somewhat happy |        953       37.64       64.93
       Very happy |        879       34.72       99.64
               .s |          9        0.36      100.00
------------------+-----------------------------------
            Total |      2,532      100.00
```

The new frequency distribution shows that 9 cases were skipped out of this question about body image. But you can see that the percentages in each category do not change substantially (generally less than a few hundredths of a percent) when they are calculated using these 9 cases.

To more clearly see the frequency order of the categories, the -sort- option can be invoked. Type tab body, sort into the Command window and press **Enter**. The table is now displayed as follows:

```
   (body_w3) P:3. |
  In general, how |
  happy or unhappy |
     are you with |
    your body and |
           physic |      Freq.     Percent        Cum.
------------------+-----------------------------------
    Somewhat happy |        953       37.77       37.77
        Very happy |        879       34.84       72.61
 Somewhat unhappy |        389       15.42       88.03
           Neither |        234        9.27       97.30
     Very unhappy |         68        2.70      100.00
------------------+-----------------------------------
            Total |      2,523      100.00
```

The logical ordering of the categories is now replaced by the frequency order, with the most prevalent category listed first and the least frequent category last. This table shows even more clearly that the most popular perception among young adults is being generally happy with their body. The cumulative percentage column indicates that more than 72% of young adults report being either somewhat or very happy.

For this particular project, you might predict that there would be a relationship between being sad and perception of one's body. The NSYR asked, "How frequently do you feel sad?" Before you examine any relationship between the two variables, it is helpful to know the distribution of both variables involved. You can use the -tab- command as above to produce this distribution for the new sad variable. If you knew that you were interested in the distribution of both variables, however, you might try listing both variables after the -tab- command. If you use the -tab- command with two variables, however, a cross-tabulation will be displayed rather than separate frequency distributions. Stata offers a slightly different command that produces multiple frequency distributions using one command line. The -tab1- command allows for several variables to be entered at one time, and separate frequency distributions for each variable are displayed, in the order in which the variables are entered in the command line.

For example, type tab1 body sad into the Command window and press **Enter**. The frequency distributions of the body and then the sad variables are displayed. -tab1- accepts all the options that have been discussed with the -tab- command, but when you invoke any option with -tab1-, those options are applied to every variable that is listed in the command line.

A common aspect of research projects is to examine patterns of behavior within particular subgroups. For example, you might only be interested in the body image of young adult females. Often, people new to quantitative analyses are tempted to eliminate the subgroup that they are not interested in (e.g., males) from the data set entirely. This strategy is not advisable because it might be important to at least initially examine patterns for the whole sample or compare females with males, even if the ultimate research goals only concern females. Therefore, rather than dropping all the males from the data, it is more effective to use `-if-` statements along with the analytic commands. If you wanted to see the distribution of the `body` variable for females only, you need to add an `-if-` statement, using the `gender` variable, to the end of the command line. The command would read `tab body if gender==#`. Remember a double equal sign is needed whenever you are asking Stata to evaluate if a statement is true.

Before you complete the command, you need to know the numeric value that is used to represent females in the `gender` variable. There are several ways to obtain this information, but perhaps the most straightforward is to perform two separate `-tab-` commands of the `gender` variable, one that invokes the `-nol-` option and one that does not. First type `tab gender` in the Command line and press **Enter**, and then type `tab gender, nol` in the Command line and press **Enter**. (If the variable of interest does not have value labels attached, the second `-tab-` command would not be necessary.) Doing so produces the following results:

```
. tab gender

(gender_w3) |
 Respondent |
     gender |      Freq.     Percent        Cum.
------------+-----------------------------------
       Male |      1,232       48.66       48.66
     Female |      1,300       51.34      100.00
------------+-----------------------------------
      Total |      2,532      100.00

. tab gender, nol

(gender_w3) |
 Respondent |
     gender |      Freq.     Percent        Cum.
------------+-----------------------------------
          0 |      1,232       48.66       48.66
          1 |      1,300       51.34      100.00
------------+-----------------------------------
      Total |      2,532      100.00
```

Now that you know females are coded as 1 on the gender variable, you can produce the frequency distribution of the body variable for females only. In this case, you might also consider using the -sort- option to see if the frequency order for females is similar to that of the entire sample. The command to produce this frequency distribution is tab body if gender==1, sort. Notice that the -if- statement comes before the option. This ordering is true for all commands. -if- statements always precede the comma that separates the command from the options. Once you type that command line into the Command window and press **Enter**, the following results are displayed:

```
   (body_w3) P:3. |
  In general, how |
 happy or unhappy |
    are you with  |
   your body and  |
           physic |      Freq.     Percent        Cum.
------------------+-----------------------------------
    Somewhat happy |        511       39.43       39.43
        Very happy |        375       28.94       68.36
 Somewhat unhappy |        246       18.98       87.35
          Neither |        117        9.03       96.37
     Very unhappy |         47        3.63      100.00
------------------+-----------------------------------
            Total |      1,296      100.00
```

A Closer Look: Displaying Numeric Codes and Value Labels

The two-command procedure for identifying the numeric codes behind a variable with value labels attached can seem cumbersome. Stata has a command that can eliminate the need to produce two tables. The -numlabel- command followed by a variable or variables and the -add- option alters that variable's value labels to include the numeric code and the value label to be displayed. For example, type numlabel gender, add into the Command window and press **Enter**. Then produce a distribution table for the gender variable (tab gender). The results are displayed as follows:

```
 (gender_w3) |
  Respondent |
```

(Continued)

(Continued)

```
     gender |      Freq.     Percent        Cum.
------------+-----------------------------------
    0. Male |      1,232       48.66       48.66
  1. Female |      1,300       51.34      100.00
------------+-----------------------------------
      Total |      2,532      100.00
```

Now it is clear that Male is coded as 0, and Female is coded as 1.

Multiple variables can be typed in a single command line and all their value labels will be changed. An even quicker way to add the numeric codes to the value labels in this way is to use the `_all` variable name. Typing `_all` into the portion of a command that accepts a list of variables executes the command on every variable in the current data set. For example, typing `sum _all` in the Command window and pressing **Enter** would produce summary statistics for every variable in the data set. The `-numlabel-` command operates similarly. Type `numlabel _all, add` into the Command window and press **Enter**. Now produce a distribution table of any variable in the data set, and it will be displayed similar to the table for the `gender` variable above.

One drawback to this method is that there are situations where you might not want the numeric codes displayed alongside the value labels. For instance, graphs can use value labels to identify portions of the chart (i.e., bars). Including the numeric code may hinder the readability of the graphs. Therefore, to remove the numeric codes from the value labels, use the `-remove-` option along with the `-numlabel-` command. Type `numlabel gender, remove` into the Command window and press **Enter**. Now when you produce a distribution table of this variable, only "Male" and "Female" are displayed as value labels.

This distribution is calculated only using female respondents. The total number of females included in this distribution is 1,296, meaning that 4 females are missing on this question (because the distribution table of `gender` showed 1,300 females present in the entire sample). Both the individual category percentages and cumulative percentages use this total in the denominator of the calculation. The results show that the majority of young adult females, 68.36%, are somewhat or very happy with their body.

HISTOGRAMS, BAR GRAPHS, AND PIE CHARTS

In addition to examining the actual frequency distribution, it can be helpful to see the distribution of the variables visually. To do so, there are three primary options: histograms, bar graphs, and pie charts. In practice, histograms and bar graphs are essentially the same, as both use bars to illustrate the relative frequency of categories within a variable. In Stata, it is easier to produce these types of graphs using the `-histogram-` command, and point-and-click box, than the bar graph command. The latter will be used for examining the relationship between two variables.

Thus far, this book has focused almost entirely on running procedures by using the Command window interface. Perhaps the one area that the point-and-click method has advantages that outweigh the Command window method is in preparing graphs. Due to the vast number of manipulations that can be made to the display of graphs, the commands to produce them can become lengthy and cumbersome. Therefore, for the sections explaining the productions of graphs, the point-and-click interface is primarily used.[1]

To use a histogram to examine the distribution of young adults' body image, start by clicking on the **Graphics** menu button at the top of the Stata window and then select **Histogram**. A window like the one below will appear:

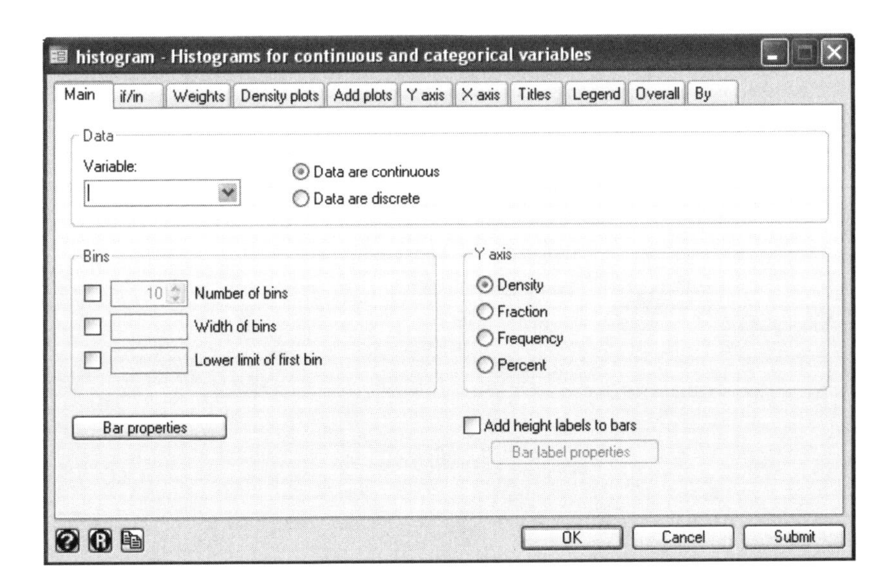

[1]As with all operations performed using the point-and-click interface, the actual command to produce a given graph is displayed in the Results window. If a user feels more comfortable with the Command window interface, it is possible to follow the point-and-click steps described and then use the displayed command to learn the necessary commands and options to produce the graphs via the Command window.

Once this window appears, you need to specify the variable that you want to produce a histogram of. Place your cursor in the Variable box in the upper left-hand corner. You can use the drop-down menu to find the `body` variable or you can simply type `body` into the box. Next, you can select what scale the Y-axis should depict. The default is set to Density, but it is more common to display either the Frequency or the Percentage. The shape of the histogram will not change regardless of which scale is chosen, but the interpretability of the Y-axis is typically enhanced by using the Percentage option. Simply click the radio button next to Percentage to set the Y-axis scale. This window also contains an option to add a "height label" to the bars. This option will display the actual number, depending on the Y-axis scale, that is associated with each category. For example, 2.70% would be listed above the "very unhappy" bar because 2.7% of the sample report being very unhappy with their body. Finally, you can select the Bar Properties button to alter the appearance (e.g., color and outline pattern) of the bars themselves.

At this point, if you pressed the OK button, a histogram would be produced. But, by default, histograms label the bars with the numeric values of the categories rather than the value labels. To alter the labeling, in order to ease the readability of the graph, click on the X axis tab while still in the histogram window. The following window will appear:

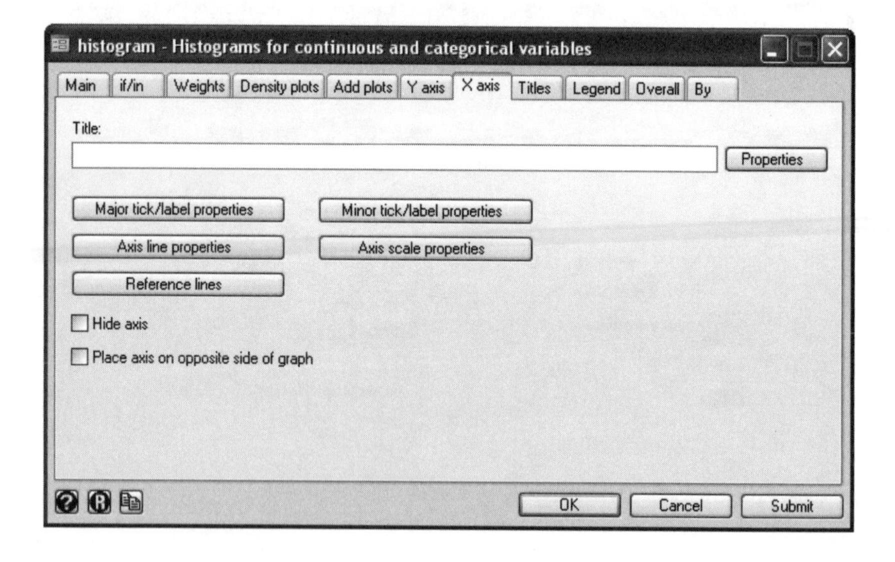

This window contains all the options for the X-axis. For example, you could enter a title, such as "Happiness With Body and Physical Appearance" in the Title box. To have the bars labeled with the appropriate value labels, press the Major tick/label properties button. Once the new window appears, select the Labels tab, which will display the following window:

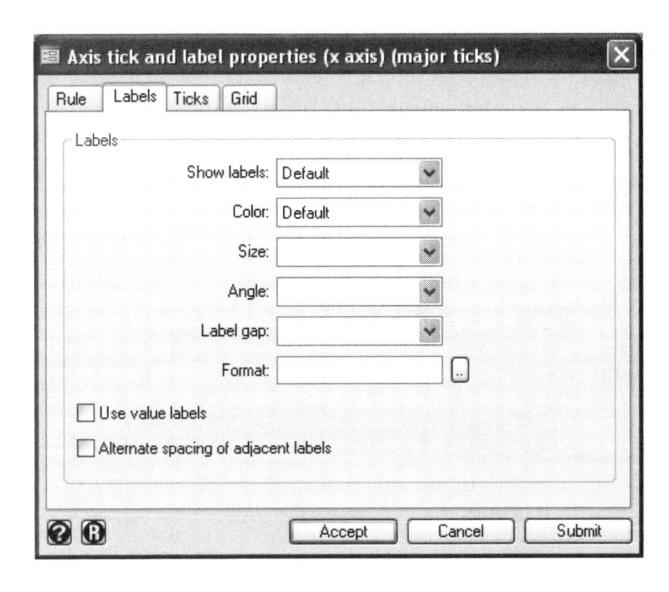

Within this window check the radio button next to Use value labels, then click Accept, and finally click OK. The resulting histogram should look like the one below:

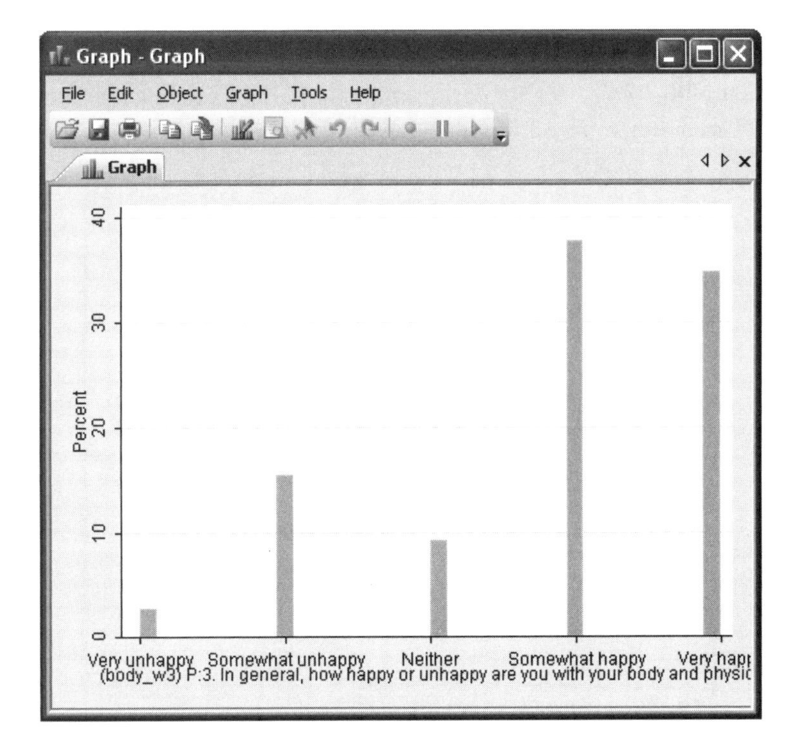

This graph can be saved by clicking on the File menu and then Save As. You can change the file format into several different picture file types, although by default it saves as a Stata graph (.gph). This graph again clearly shows that the two happy categories are far more prevelant among young adults than are the unhappy categories. Notice, however, that the Y-axis is scaled to the maximum percentage of the given categories, not 100%.

To alter the display of the Y-axis, click on the Graphics menu and then Histogram. All the options you have previously set are still in the Histogram window. If you wanted to clear all those options, you can press the R icon located in the lower left corner. Select the Y axis tab, followed by the Axis scale properties button.

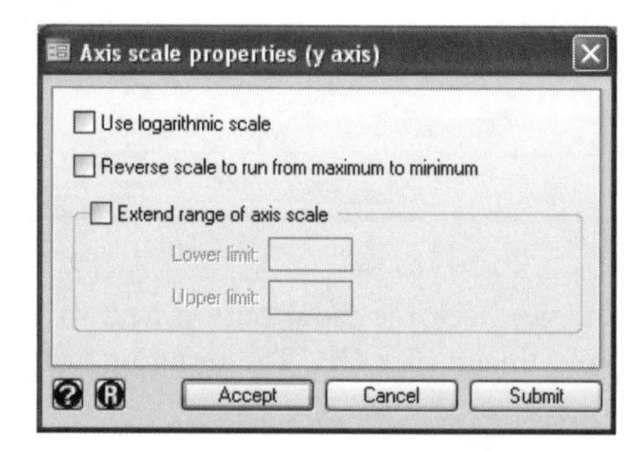

In the window that appears, select the Extend range of axis scale button, set the Lower Limit to 0 and the Upper Limit to 100 and click Accept. Now select the Major tick/label properties button followed by the Suggest # of ticks button to reveal the following window:

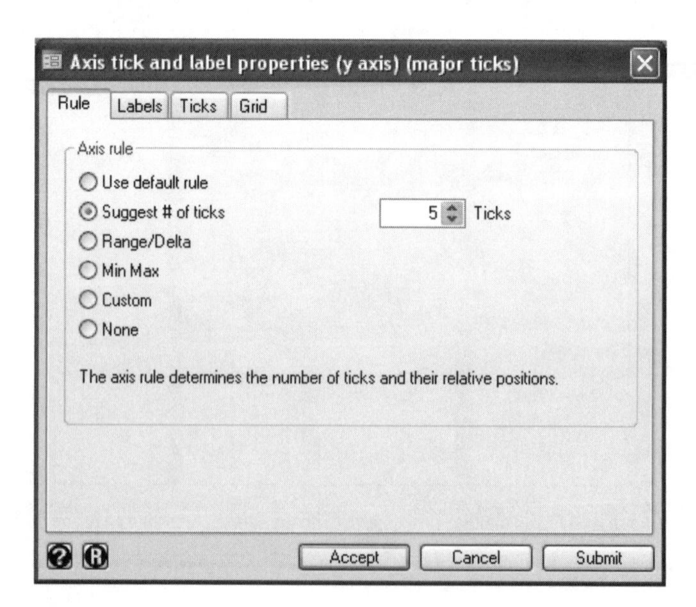

Finally, change the Ticks value to 10, instead of 5, click Accept and then OK. The histogram should now look as follows:

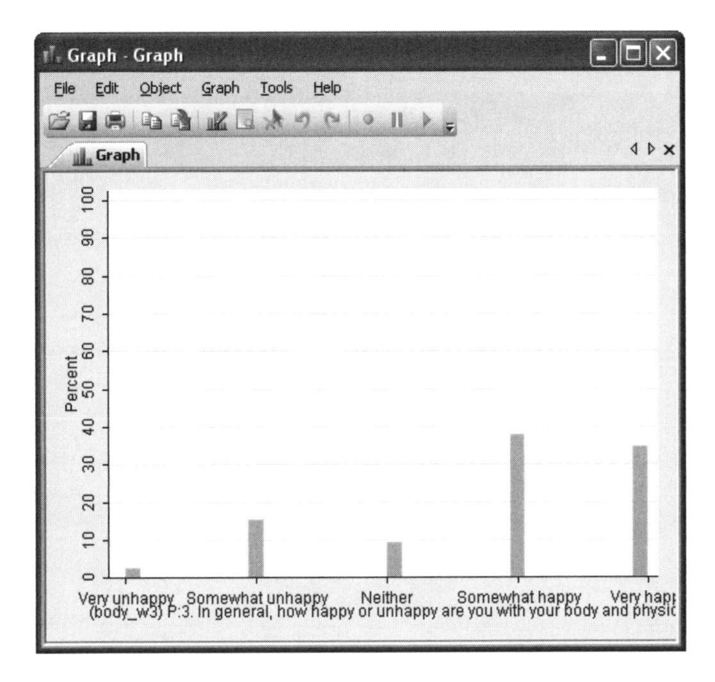

The Y-axis is now on a 0% to 100% scale, providing a more accurate depiction of the absolute percentages of each category.

If you are interested in the relative percentage of each category, a Pie Chart may be a better graphic display option. To produce a pie chart, click on the Graphics menu and then the Pie chart option. The following window will be displayed:

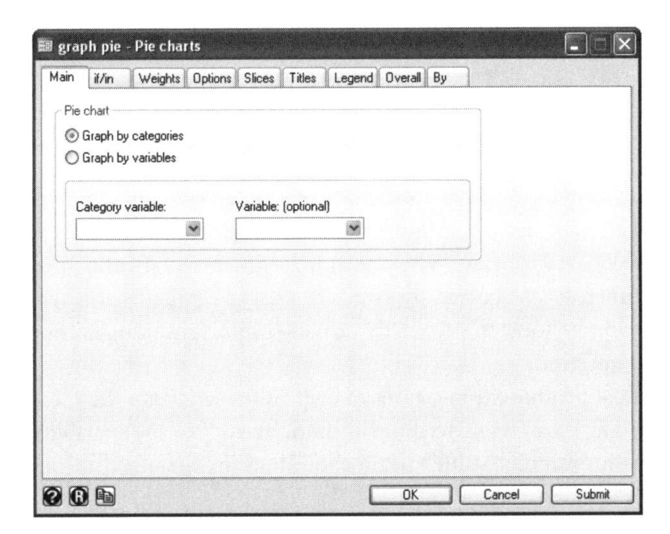

Select the **Category variable** box and type `body` (or select it from the drop-down menu). Finally click **OK** and the Pie Chart should look like the following:

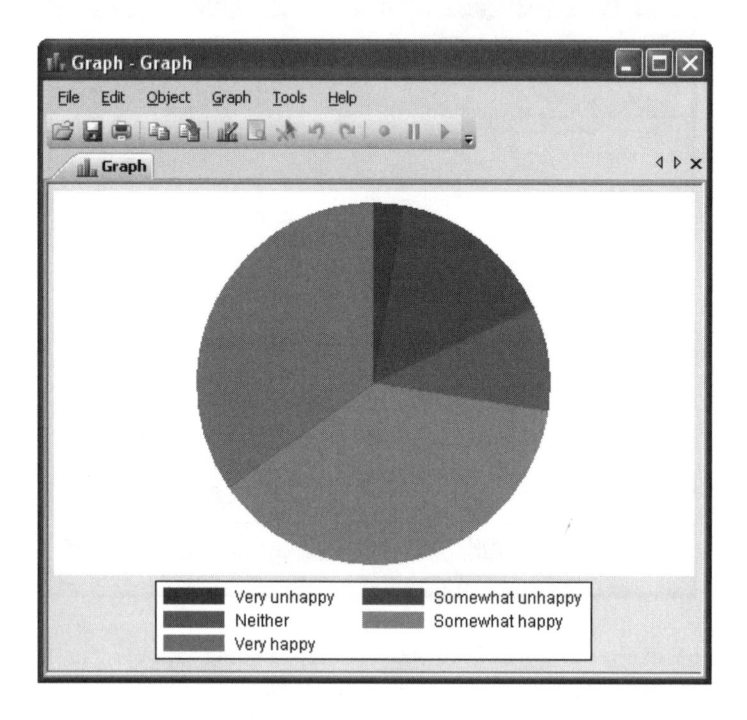

As with the histogram menu, there are several options that can be selected to alter the appearance of a Pie Chart. This basic display provides an effective picture of the distribution of the categories.

Measures of Central Tendency and Variability

A frequency distribution is a good first step in describing a variable. Such distributions are especially helpful for nominal or ordinal measures that do not have numerous categories. When working with interval-ratio measures that contain numerous categories, measures of central tendency and variability are more appropriate tools to succinctly provide important information about the variable in question.

The most prominent measures of central tendency are the mean, median, and mode. The mean refers to the arithmetic average of the variable's value, the median is the "middle" value (when the values are arranged in numerical or value order), and the mode is the most prevalent category. The range, interquartile range (IQR), variance, and standard deviation are the most commonly reported measures of variability. The range represents the difference between the maximum and the minimum value, and the IQR is the difference between the 25th and 75th percentile values. The variance is the "average" difference between each case's value and the mean value, while the standard deviation is the square root of the variance.

You actually have already learned how to obtain one of these measures. The `-tab-` command along with the `-sort-` option is the quickest way to obtain the mode. The category that is listed first, when the `-sort-` option is used (or the category with the highest percentage if it is not), is the modal category or value.

As introduced in Chapter 2, the `-sum-` command (short for `-summary-`) displays several of the other necessary measures. In addition to being interested in young adults' perceptions of their bodies, a researcher might be concerned with their actual body dimensions. One variable in the NSYR data that assesses body type is `bmi`. This variable represents the body mass index, or BMI, of the respondents. According to the National Institute of Health, BMI values greater than 30 indicate that a person is obese, values between 25 and 29.9 indicate that a person is overweight, and values 18.5 through 24.9 represent a person who has normal weight range.[2] To calculate BMI, each person's exact height and weight are used, resulting in an interval-ratio measure.

Type `sum bmi` into the Command window and press **Enter**. The following results are displayed:

```
Variable |       Obs        Mean    Std. Dev.       Min        Max
---------+--------------------------------------------------------
     bmi |      2509    24.73744    5.141812   14.01495   63.49296
```

Five figures are presented with the default `-sum-` command. First, the number of valid observations (i.e., nonmissing) is shown. For the `bmi` variable, 23 cases are missing (2,532 – 2,509 = 23). Most likely several people did not wish to report their weight, resulting in a missing `bmi` value. Next, the mean is reported. A value of 24.74 suggests that, on average, young adults are at the very top of the normal weight range. Next, a standard deviation of 5.14 suggests a reasonably wide distribution of values. The minimum and maximum values are displayed, making it easy to calculate the range by subtracting the former from the latter. For the `bmi` variable, there is a range of almost 50 units (63.49 – 14.01 = 49.48).

[2] http://www.nhlbisupport.com/bmi/

The default -sum- command does not list the median, IQR, or the variance. To produce these statistics, the -detail- option must be invoked.

Type `sum bmi, detail` in the Command window and press **Enter** to produce the following results:

```
          (bmi_w3) Body Mass Index (NIH calculation)
-------------------------------------------------------------
      Percentiles      Smallest
 1%      16.9512       14.01495
 5%     18.75257       14.22837
10%     19.57563       14.64583       Obs                2509
25%     21.28223       14.76581       Sum of Wgt.        2509

50%     23.62529                      Mean           24.73744
                        Largest       Std. Dev.      5.141812
75%     26.95946       52.99345
90%     31.32101       53.21151       Variance       26.43823
95%     35.03738       56.48531       Skewness       1.536935
99%     41.97015       63.49296       Kurtosis       7.081053
```

The variance is now displayed in the lower right, and for the bmi variable it is 26.44. The median, however, is not definitively labeled. Instead, Stata lists the major percentiles. The 50% value, or the 50th percentile, is equivalent to the median value. For bmi, the median and mean are relatively similar, suggesting a normal distribution of values. The IQR can also be easily calculated by subtracting the 25th percentile value (21.18) from the 75th percentile value (26.95), producing a figure of 5.68. In addition to numerous percentiles, the -detail- option also lists the Skewness and Kurtosis scores, which are two additional measures of variability.

With or without the -detail- option, the -sum- command can handle multiple variables listed in a single command line. For example, you could type `sum bmi agecats,detail` into the Command window and press **Enter**. The detailed summary statistic results would be presented, one after the other, for both variables. One drawback of this method is that it is difficult to quickly see similar figures (e.g., the mean) across several variables, and the -detail- option may list several statistics that are not needed.

To present a more consolidated table of the measures of central tendency and variability across numerous variables *and* control the statistics that are displayed, the -tabstat- command is an effective alternative. By default without any options, the -tabstat- command only reports the mean. Displaying additional statistics is controlled by the -statistics(statname)- option (shortened -stat(statname)-). The "statname" in this option refers to a particular code for the statistic you would like to be presented. As with most of the Stata commands you have encountered, the codes for each statistic is intuitively straightforward (e.g., the code for the mean is `mean`). See the "A Closer Look" box for a full list of the available statistics and their codes.

A Closer Look: Statistics and Their Codes for Use With tabstat, stat (statname)

The following table lists all the statistics that can be displayed with the `-tabstat-` command and their associated codes that are typed in the `-stat(statname)-` option.

Statistic	**statname** Code
mean	mean
count of nonmissing observations	count
same as count	n
sum	sum
maximum	max
minimum	min
range = max − min	range
standard deviation	sd
variance	variance
coefficient of variation (sd/mean)	cv
standard error of mean (sd/sqrt(n))	semean
skewness	skewness
kurtosis	kurtosis
1st percentile	p1
5th percentile	p5
10th percentile	p10

(Continued)

(Continued)

Statistic	statname Code
25th percentile	p25
median (same as p50)	median
50th percentile (same as median)	p50
75th percentile	p75
90th percentile	p90
95th percentile	p95
99th percentile	p99
interquartile range = p75 - p25	iqr
equivalent to specifying p25 p50 p75	q

Source: Table courtesy of *Stata Manual.*

To present the set of measures of central tendency and variability listed at the beginning of this section for both the `bmi` and `agecats` variable, type `tabstat bmi agecats, stat(mean median min max range iqr variance sd)`. The following results are displayed:

```
   stats |       bmi   agecats
---------+--------------------
    mean |  24.73744  20.01817
     p50 |  23.62529        20
     min |  14.01495        17
     max |  63.49296        24
   range |  49.47801         7
     iqr |  5.677233         2
variance |  26.43823  2.088963
      sd |  5.141812  1.445324
------------------------------
```

The same values for the `bmi` variable are listed as when the `-sum-` command was used, but the `-tabstat-` command presents them in a more concise fashion. It is also much easier to see the pertinent figures across the two variables. As when using the -detail- option, the median is labeled by the 50th percentile or p50. The standard deviation is given by sd.

A potential problem arises when using the -tabstat- command in this manner when several variables are entered at one time. Eventually, the display becomes too wide to fit on the screen. In this case, it may be more effective to also invoke the -columns(variables/statistics)- option (shortened -col(var/stat)-). When you do not use this option, it is automatically set as -columns(variables)-, and the variables are organized in the columns, as shown above.

If you type tabstat bmi agecats, stat(mean median min max range iqr variance sd) col(stat) in the Command window (remember you can use the **Page Up** function, the Review window, or a do file to eliminate retyping this lengthy command) and press **Enter**, the results are now displayed as follows:

```
variable |      mean       p50       min       max     range       iqr  variance        sd
---------+--------------------------------------------------------------------------------
     bmi |  24.73744  23.62529  14.01495  63.49296  49.47801  5.677233  26.43823  5.141812
 agecats |  20.01817        20        17        24         7         2  2.088963  1.445324
---------+--------------------------------------------------------------------------------
```

You can see that this modified display would be even more effective if you were examining a large number of variables at once.

BOX PLOTS

A box plot is a graphical presentation of all the information displayed in the -tabstat- table above. This type of plot is especially useful for examining the overall distribution of an interval-ratio variable and determining whether that distribution is skewed or affected by outliers.

A box plot is one graph that is almost as easy to produce using the Command window as it is through the point-and-click method. The command is -graph box-, followed by the variable name(s) that you want to produce a box plot of. It is possible to enter multiple variables at one time, and all of the box plots will be presented on one graph. However, this type of box plot should only be used when the variables are measured with similar units. For example, it would not make a great deal of sense to place the box plots for bmi and agecats side-by-side because the units are not equivalent.

To produce a box plot of `bmi` using the Command window, type `graph box bmi` in the Command window and press Enter. The graph should display like the one below:

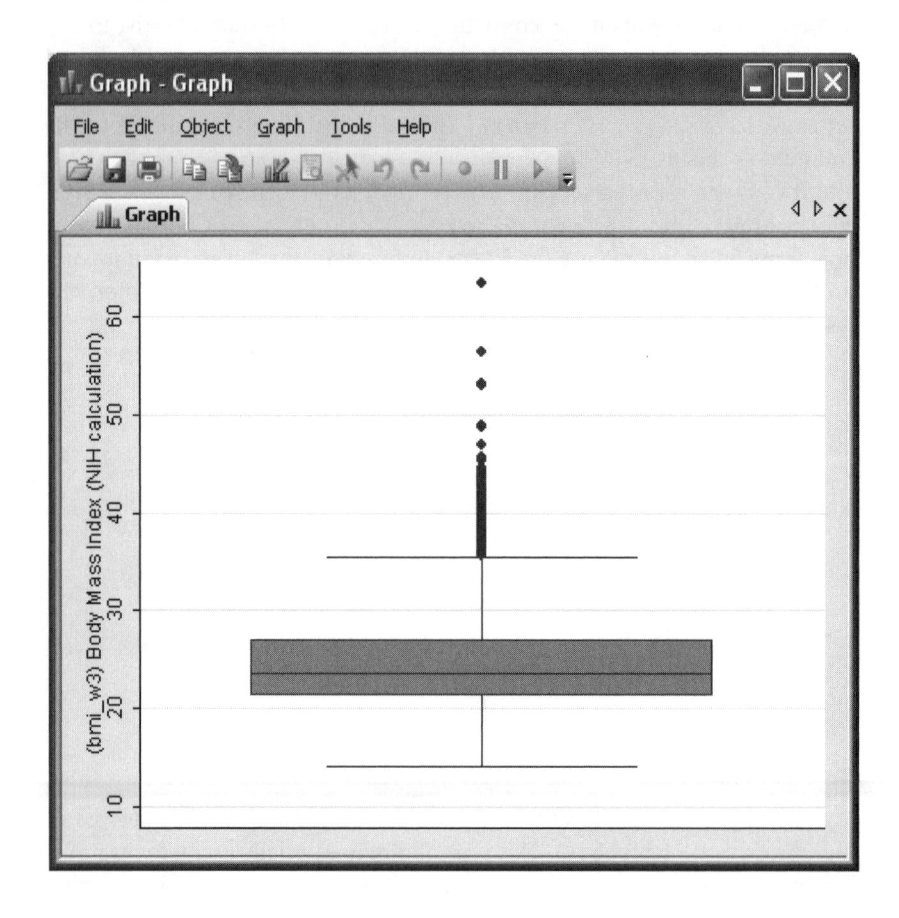

The box plot of `bmi` shows that the distribution is relatively normal, as the line indicating the median is close to the center of the box, and the box itself is approximately in the middle of the range. The plot further shows that outliers may be a concern as there are several extreme values, represented by the dots, at the high end of the distribution.

Producing a similar plot using the point-and-click interface is similarly straightforward. Click on the Graphics menu and then Box Plot. The following window will appear:

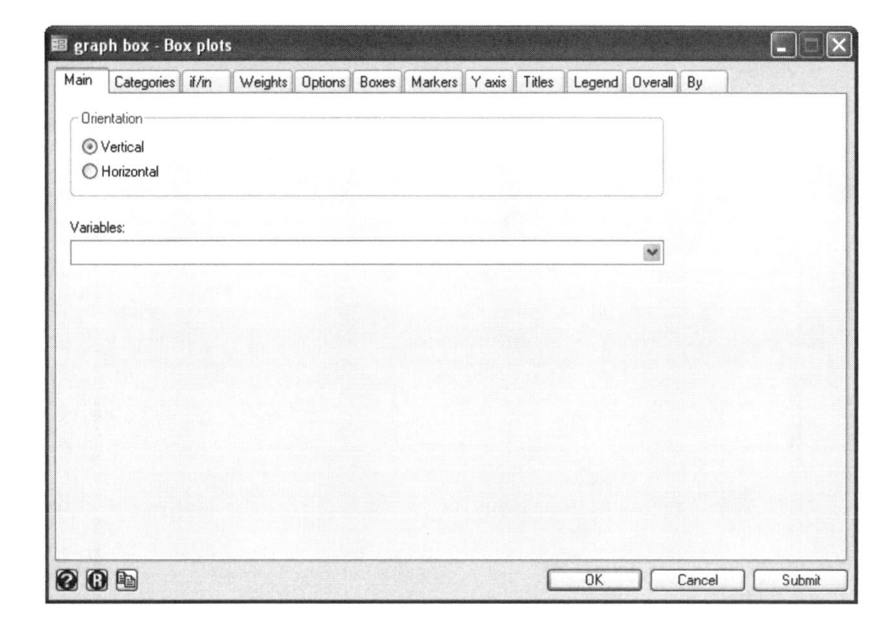

In the Variables box, either use the drop-down menu to find or type in `bmi`, and then click OK. The same box plot shown above will be displayed.

A Closer Look: Using Histograms to Examine Central Tendency and Variability

Creating histograms using the point-and-click method was discussed earlier, but these graphs are also useful for quickly examining the distribution of interval-ratio variables. Furthermore, because interval-ratio variables do not use value labels (i.e., their values are their labels), these graphs are easier to produce using the Command window interface when examining such variables.

The command to produce a histogram is `-histogram-` (shortened `-hist-`). To quickly produce a histogram of the `bmi` variable, you could type `hist bmi` in the Command window. But, as with the point-and-click method, the default Y-axis scale is each category's density. This scale can be changed by invoking either the `-percent-` or `-frequency-` (abbreviated `-freq-`) option.

(Continued)

(Continued)

For example, type `hist bmi, freq` into the Command window and press **Enter**. The following graph will be displayed:

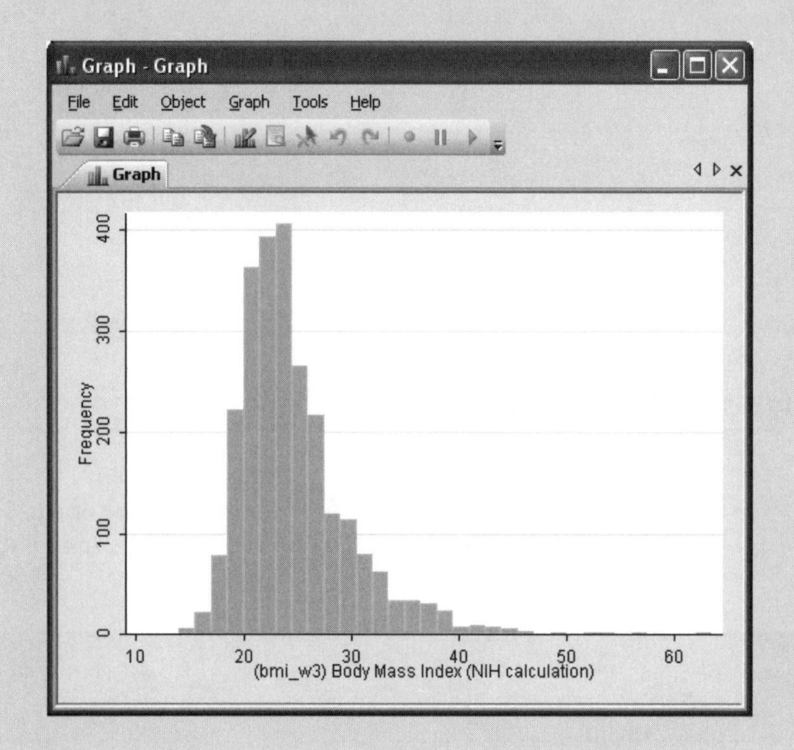

As you can see, the histogram very clearly illustrates the shape of the distribution. Similar to the other tools that have been used, the graph shows that the distribution is essentially normal with a few outliers at the high end of the distribution.

Summary of Commands Used in This Chapter

```
*Frequency Distributions
tab body
tab body, mis
tab body, sort
tab1 body sad
```

```
tab gender
tab gender, nol

tab body if gender==1, sort


*Measures of Central Tendency and Variability
sum bmi
sum bmi, detail
tabstat bmi agecats, stat(mean median min max range
iqr variance sd)
tabstat bmi agecats, stat(mean median min max range
iqr variance sd) col(stat)


*Box Plots
graph box bmi

*A Closer Look: Displaying Numeric Codes and Value Labels
numlabel gender, add
tab gender

numlabel _all, add

numlabel gender, remove
tab gender

*A Closer Look: Using Histograms to Examine Central
Tendency and Variability
hist bmi, freq
```

Exercises

Use the original `Chapter 4 Data.dta` for the following problems. [Optional: Complete the exercises using a do file and save the results using a log file. See Chapter 3 for an explanation of how to use these files.]

1. Produce a frequency distribution of how important religion is to young adults (`faith1`).
2. Display another frequency distribution of the `faith1` variable, including missing categories, that is arranged with the most frequent category displayed first.

3. Produce a frequency distribution for both how important religion is to young adults (`faith1`) and how much young adults care about the elderly (`crelder`) with one command.
4. Produce a frequency distribution for the variable `crelder` for respondents who think religion is extremely or very important.
5. Produce a percentage histogram of the `faith1` variable that uses value labels and has a Y-axis that ranges from 0 to 100.
6. Make a pie chart of the `crelder` variable.
7. Show the detailed measures of central tendency and variability for the number of children young adults desire (`kidwntmn`).
8. Produce a table, with the variables in the rows, that displays the mean, median, standard deviation, and variance for the `kidwntmn` variable and the number of religious retreats young adults have attended (`relretrt`).
9. Generate a box plot for the `kidwntmn` variable.
10. Produce a frequency histogram (using the Command window) of the `relretrt` variable.

5

Relationships Between Nominal and Ordinal Variables

After examining the distributions and descriptive statistics for individual variables, the next step in most research projects is to investigate the relationship between two or more variables. There are countless techniques for assessing these types of relationships, making it difficult for new analysts to know where to start. Perhaps the best way to narrow down the possible options is to identify the level of measurement of the two variables in question. Most statistical techniques are designed to be used with a specific type of variable.

This chapter focuses on strategies for inspecting the relationship between nominal or ordinal variables. Nominal and ordinal variables both have a limited number of categories and, usually, do not correspond inherently to a numeric value. Examples of nominal and ordinal variables include race, employment status, level of agreement on a Likert-type scale, and denomination affiliation.

All the examples that follow use the `Chapter 5 Data.dta`, available at **www.sagepub.com/longest**. This data set includes a more extensive set of variables from the National Study of Youth and Religion (NSYR) data and contains the full Wave 3 sample of 2,532 young adults. All the missing cases have been replaced with appropriate missing codes in this data set, with `.d` referring to a response of "Don't Know," `.r` to a response of "Refused," and `.s` to a case that was legitimately skipped out of a question based on the survey design (i.e., skip pattern).

Cross-Tabulations

One of the best ways to examine how two nominal or ordinal variables are related is by creating a cross-tabulation. A cross-tabulation shows how the distribution of one variable fits within the categories of another variable. For example, the NSYR asked respondents to indicate how much they agree with the statement "You like to take risks." Chapter 4 described how to examine the distribution of a single variable, but you might predict that level of agreement with risk taking would vary across education levels. A logical hypothesis would be that young adults who have had more education are more averse to risk than are young adults who have received less education. To test this prediction, you could start by creating a table that displays the distribution of agreement for young adults who have attended college next to the distribution of agreement for young adults with no college experience. That is, you would construct a cross-tabulation.

As discussed previously, however, it is always a good strategy to look at the individual distributions before investigating a relationship between two variables. The variable for the risk question is named `risks`. The NSYR also contains a variable called `cu_attco`, which is an indicator of whether the respondent has ever attended any college.

Type `tab1 risks cu_attco` into the Command window and press **Enter**. Note that `-tab1-` is used, rather than `-tab-`, to produce multiple, individual frequency distributions in one command line. Both of the following frequency distributions are displayed:

```
-> tabulation of risks

     (risks_w3) P:18. You like to take |
  risks. (Do you strongly agree, agree, |
                             disagree, |      Freq.     Percent        Cum.
---------------------------------------+-----------------------------------
                        Strongly agree |        444       17.61       17.61
                                 Agree |      1,389       55.10       72.71
Undecided/DK (Interviewer: Do not read) |         13        0.52       73.22
                              Disagree |        593       23.52       96.75
                     Strongly disagree |         82        3.25      100.00
---------------------------------------+-----------------------------------
                                 Total |      2,521      100.00

-> tabulation of cu_attco

(cu_attendc |
    oll_13) |
       Ever |
   attended |
    college |      Freq.     Percent        Cum.
------------+-----------------------------------
         No |        839       33.14       33.14
        Yes |      1,693       66.86      100.00
------------+-----------------------------------
      Total |      2,532      100.00
```

The second table shows that almost 67% of the sample has attended some college, and all the cases have responses on this education variable. The `risks` variable, however, is a bit more problematic. First, the cumulative percentage column shows that a majority of young adults (72.71%) either agree or strongly agree that they like to take risks. Conversely, only 3.25% strongly disagree with liking to take risks. Finally, only 13 cases (0.52%) said they were unsure about how much they liked to take risks. In certain analyses, you could consider "undecided" its own unique, meaningful category that deserves examination (i.e., what makes some young adults unsure about their risk taking). But 13 cases are probably not enough to conduct a valid analysis.

The `risks` variable is an illustration of a common occurrence in quantitative research—collecting the data is only half the battle. It is rare for data to be ready-made for analysis. Typically you need to do some data management to prepare variables for the type of analysis you are attempting to conduct.

There are several ways that you could reconfigure the `risks` variable to handle the issues noted. The original research question, however, was whether college attendees and nonattendees differ in whether they like to take risks or not. Therefore, it makes sense to turn the `risks` variable into a dichotomous indicator of the respondent agreeing with the statement about liking to take risks. Defining the variable in this way makes it more defensible to code the "undecided" respondents into the nonagreement category, as opposed to setting them as missing. To create this variable, all respondents who are in the strongly agree or agree categories should be coded into one category, and the respondents in the undecided, disagree, and strongly disagree categories should be coded into a different category.

The `-recode-` command, along with its `-gen(newvar)-` option, is the most effective strategy to use in this situation. Before you type the `-recode-` command, you need to know the numeric codes of the categories of the `risks` variable. To see these values, type `tab risks, nol` in the Command window and press **Enter**.

```
   (risks_w3) |
    P:18. You |
      like to |
 take risks. |
      (Do you |
     strongly |
       agree, |
       agree, |
```

(Continued)

(Continued)

```
   disagree, |      Freq.     Percent        Cum.
-------------+-----------------------------------
           1 |        444       17.61       17.61
           2 |      1,389       55.10       72.71
           3 |         13        0.52       73.22
           4 |        593       23.52       96.75
           5 |         82        3.25      100.00
-------------+-----------------------------------
       Total |      2,521      100.00
```

This table now makes it easier to construct the `-recode-` command. Remember, it is helpful to state what you want Stata to do in a way that you would explain it to a smart colleague, and then take that verbal statement and turn it into the appropriate code. For the example, you might think about saying "turn Categories 1 and 2 into 1, turn Categories 3 through 5 into 0, and then create a new variable to hold this changed categorization."

Using this statement produces the code: `recode risks (1/2=1) (3/5=0), gen(agrisk)`. Type this command into the Command window and press **Enter**. Remember you could name the new variable anything you want. `agrisk` is a way to clearly indicate that the variable is an indicator of agreeing with the risk question. Similarly, you could use any two numbers you want to represent each category. Assigning 0 and 1 to these types of indicator variables, however, is a common practice. Next, to check that the command did what you were intending, type `tab risks agrisks` into the Command window and press **Enter**.

```
                     |     RECODE of risks
 (risks_w3) P:18. You | ((risks_w3) P:18. You
  like to take risks. |  like to take risks.
     (Do you strongly |    (Do you strongly
        agree, agree, |         agree
           disagree, |         0          1 |     Total
----------------------+----------------------+----------
       Strongly agree |         0        444 |       444
                Agree |         0      1,389 |     1,389
Undecided/DK (Intervi |        13          0 |        13
             Disagree |       593          0 |       593
    Strongly disagree |        82          0 |        82
----------------------+----------------------+----------
                Total |       688      1,833 |     2,521
```

The results indicate that the `-recode-` command categorized the cases correctly, but the new variable does not have any value labels. Right now, it is

A Closer Look: Recoding for Direction

The `risks` variable illustrates another common issue when using secondary data. The variable is coded "backward" from the way most people would think about the measure. Notice that higher numeric values (i.e., categories listed at the bottom of the frequency distribution) are associated with more *disagreement*. This variable is intended to measure how much someone agrees with the statement about liking to take risks, suggesting that higher values should be associated with stronger agreement. This type of reverse coding can occur when surveys flip response options to avoid participants falling into a pattern of responding with the same category without thinking about the question.

This coding scheme does not affect the substantive results of an analysis, but it can make the findings more difficult to interpret. For instance, if the original `risks` variable was used as a dependent variable, any "positive" relationship would actually mean that the independent variable caused more disagreement with liking to take risks. This type of finding is confusing and can even lead to incorrect interpretations. Therefore, it is typically easier to simply change the values of the variable so that they align with the more straightforward meaning of the variable (i.e., higher values indicate higher agreement).

To perform this type of recategorization, the `-recode-` command with the `-gen(newvar)-` option is the most effective strategy. The most difficult aspect of this type of recoding is keeping the old and new values aligned. Using the original distribution tables (shown in the text) and thinking about what you want the command to do will make the process easier. You are going to ask Stata to make anyone who responded with "Strongly agree" (coded as 1 on the original variable) to equal 5 (i.e., the highest possible value) on the new variable. Similarly, anyone who is 2 ("Agree") needs to be 4, all the 3s (i.e., Undecided) can stay the same, 4s ("Disagree") should become 2s, and 5s ("Strongly disagree") should be coded as the lowest category of 1 on the new variable. Fortunately, you do not need to worry about the ordering of this transformation (i.e., there is no need to be concerned about turning all of the 1s into 5s before the 5s are turned into 1s). Stata only needs to know the old values, and what they should equal on the new variable. Therefore, you can type the command `recode risks (1=5) (2=4)`

(Continued)

(Continued)

`(3=3) (4=2) (5=1), gen(likerisk)` in the Command window and press **Enter**.

The `(3=3)` is not technically needed as a part of the command line. Any values that are not explicitly included in the `-recode-` command line are simply copied into the new variable with the same value from the original variable. Including these types of statements can be helpful in ensuring that you are changing the values in the appropriate way and not missing any needed recodes. Because the original variable has five values, it can be helpful to include five separate recode specifications, even if one of them simply tells Stata to copy the old value.

If you produce a frequency distribution of the new variable, by typing `tab likerisk` into the Command window and pressing **Enter**, the following table is displayed:

```
  RECODE of |
      risks |
((risks_w3) |
   P:18. You |
    like to |
take risks. |
    (Do you |
   strongly |
      agree |      Freq.     Percent        Cum.
------------+-----------------------------------
          1 |         82        3.25        3.25
          2 |        593       23.52       26.78
          3 |         13        0.52       27.29
          4 |      1,389       55.10       82.39
          5 |        444       17.61      100.00
------------+-----------------------------------
      Total |      2,521      100.00
```

Comparing this new distribution to the original variable, you can see that all the frequencies and percentages have stayed the same, only the ordering has changed. Now the cases that responded in some form of agreement are coded with a larger numerical value than respondents who reported being undecided or in some form of disagreement.

easy to remember that 1 means a person agreed with the statement about risks. In a few weeks though, this distinction may get a bit fuzzier. Therefore, it is helpful to apply a value label to prevent this confusion and make future tables with this variable easier to read.

As discussed in the Data Management: Working With Labels section of Chapter 3, attaching value labels involves two steps. First define the value label. Type `lab def ynagree 0 "No Agree" 1 "SA-Agree"` into the Command window (or do file) and press **Enter**. You have defined a new value label called "`ynagree`" such that cases coded as 0 are labeled as not agreeing and cases coded as 1 are labeled as strongly agreeing or agreeing.

Next, the defined value label must be attached to the variable. Type `lab val agrisk ynagree` into the Command window (or do file) and press **Enter**.

Finally, even though Stata automatically assigns a variable label when the `-gen(newvar)-` option is used in the `-recode-` command, it may be helpful to assign your own. Type `lab var agrisk "Agree or Not with Taking Risks (rc P:18)"` into the Command window (or do file) and press **Enter**. This new variable label clearly explains what the variable means and also denotes that it was a recoded (i.e., rc) version of question P:18. Now, produce a frequency distribution of the new variable by typing `tab agrisks` into the Command window and pressing **Enter**.

```
   Agree or |
   Not with |
     Taking |
  Risks (rc |
      P:18) |      Freq.     Percent        Cum.
------------+-----------------------------------
   No Agree |        688       27.29       27.29
   SA-Agree |      1,833       72.71      100.00
------------+-----------------------------------
      Total |      2,521      100.00
```

This process may seem a bit cumbersome right now, but taking the time to adequately prepare the data in the beginning of a research project saves endless frustration in the long run. Even more important, this preparation is the type of work that is required to conduct valid and useful quantitative analysis. Although it may be quicker and easier to overlook some of the minor issues (e.g., the limited number of cases in the undecided category), the extra effort at the start always produces more effective research in the end.

Now that the data are prepared, producing the cross-tabulation is relatively straightforward. To display a cross-tabulation between the recoded risk variable and the college attendance indicator variable, type `tab agrisk cu_attco` into the Command window and press **Enter**. The following table is shown:

```
   Agree or |
   Not with |
     Taking |   (cu_attendcoll_13)
  Risks (rc | Ever attended college
     P:18)  |        No        Yes |     Total
------------+----------------------+----------
   No Agree |       187        501 |       688
   SA-Agree |       647      1,186 |     1,833
------------+----------------------+----------
      Total |       834      1,687 |     2,521
```

Notice that whichever variable is entered first after the `-tab-` portion of the command appears in the rows of the table, while the second variable is placed in the columns. There is no set rule when constructing a cross-tabulation as to which variable belongs in the columns and which belongs in the rows. It can be helpful, however, to develop a consistent system for yourself to prevent confusion. For the purposes of this book, the dependent variable (i.e., the outcome) is always placed in the rows, and the independent variable (i.e., the predictor or cause) is placed in the columns. In the current example, the prediction is that attending college influences young adults' preference for taking risks. Risk taking is the dependent variable and placed in the rows (typed immediately after `-tab-`), while college attendance is the independent variable and placed in the columns (typed as the second variable in the command line).

The table above provides initial evidence of a relationship between risk taking and college attendance. The `SA-Agree` row indicates that more college attendees than nonattendees agree that they like to take risks. The `Total` row, however, shows that there are more college attendees in the sample than nonattendees. The difference in frequencies in agreeing with risk taking between college attendees and nonattendees could stem from the fact that there are more college attendees overall. Simply because college attendees are more prevalent, they are more likely to be in the agreeing with taking risks category. Indeed, the `No Agree` line shows that more college attendees do not agree with the statement about taking risks than nonattendees. Using the frequencies alone, therefore, is not an adequate way to assess the relationship between two variables.

To address this issue, the percentages of college attendees and nonattendees in each agreement category should be compared, rather than only the frequencies. This step is where cross-tabulations can be slightly tricky. You must decide how the percentages should be calculated. That is, you must determine whether the percentages should be based on the row totals or column totals. A way to help remember where the percentages should be is to think about what you want to compare. For the current example, you are comparing college attendees and nonattendees, meaning you need the percentage of *attendees* and the percentage of *nonattendees* that agree with the statement about taking risks. Therefore, the percentages should be based in the columns because that is where the college attendance categories are.

Including percentages in cross-tabulation requires that either the `-column-` (shortened `-col-`) or `-row-` option be invoked after the `-tab-` command. Because the column percentages are needed, type `tab agrisks cu_attcol, col` in the Command window and press **Enter**. The new table will be displayed as follows:

```
+-------------------+
| Key               |
|-------------------|
|     frequency     |
| column percentage |
+-------------------+

   Agree or |
   Not with |
     Taking |   (cu_attendcoll_13)
  Risks (rc | Ever attended college
      P:18) |        No        Yes |     Total
-----------+----------------------+----------
   No Agree |       187        501 |       688
            |     22.42      29.70 |     27.29
-----------+----------------------+----------
   SA-Agree |       647      1,186 |     1,833
            |     77.58      70.30 |     72.71
-----------+----------------------+----------
      Total |       834      1,687 |     2,521
            |    100.00     100.00 |    100.00
```

This new table is very similar to the previous one, expect now the column percentages have been added, and there is a key at the top indicating what each figure represents. This table makes it easier to compare the difference between the preference for risk taking of college attendees and nonattendees. A total of 77.58% of nonattendees (647/834 = 77.58) agree or strongly agree that they like to take risks, compared with 70.30% of college attendees (1,186/1,687 = 70.30) agreeing or strongly agreeing with this statement. In other words, there is a 7.28% difference between college attendees and nonattendees in terms of their agreement with liking to take risks. This difference suggests that young adults who do not attend college are more likely to like to take risks than are young adults who have attended college.

If the row percentages were needed, the `-row-` option could either be added to the above command line or replace the `-col-` option (both can be invoked and will be shown in the same table if needed). Furthermore, if you want a more concise table, the key and the frequencies can be suppressed, using the-`nokey`- and -`nofreq`- options, respectively.

Type `tab agrisk cu_attco, col nokey nofreq` into the Command window and press **Enter**. The following, condensed table is displayed:

```
  Agree or |
  Not with |
    Taking |   (cu_attendcoll_13)
 Risks (rc | Ever attended college
    P:18) |         No        Yes |     Total
-----------+----------------------+----------
  No Agree |      22.42      29.70 |     27.29
  SA-Agree |      77.58      70.30 |     72.71
-----------+----------------------+----------
     Total |     100.00     100.00 |    100.00
```

This type of table may be ideal for use in your actual research report. It is advisable, however, to always examine the frequencies in addition to the percentages. Just as it is difficult to adequately interpret the difference between two groups using only the frequencies, it is similarly difficult to adequately compare the difference using only the percentages. For example, if you produced the simplified table above first, you might not know if one of the categories contains very few cases. If there were only 25 young adults who did not attend college in the sample, for instance, the 77.58% might be interpreted differently than when there are more than 800 nonattendees in the sample. In any research project, it is best to start by examining more information before deciding which pieces are less important and can be suppressed.

CHI-SQUARE TEST

The tables above reveal that there is a difference in the percentage of college attendees and nonattendees who agree that they like to take risks. The 7% difference indicates that nonattendees agree that they like to take risks more than college attendees. But this disparity could have occurred by chance, perhaps due to some anomalies in the sample. To further test whether this discrepancy actually occurs in the total population of young adults, a statistical test is needed. The most common test for determining the significance of a relationship between two nominal or ordinal variables is the chi-square test. Again, a full statistics text should be consulted to understand all the details of this test, but a chi-square test essentially indicates whether the observed frequencies in each cell of the table are significantly different from what the frequencies would have been if the two variables were not related. The latter frequencies are often referred to as the "expected frequencies."

If you were interested in computing the statistic by hand, Stata offers an intuitive option, `-expected-`, that displays the expected frequencies in each cell. To have this new figure displayed, type the same command as you did above and simply add the `-expected-` option. Typing `tab agrisk cu_attco, col expected` into the Command window and pressing **Enter** displays the new table:

```
+--------------------+
| Key                |
|--------------------|
|     frequency      |
| expected frequency |
| column percentage  |
+--------------------+

   Agree or |
   Not with |
     Taking |   (cu_attendcoll_13)
  Risks (rc |  Ever attended college
     P:18)  |         No        Yes |     Total
------------+-----------------------+----------
   No Agree |        187        501 |       688
            |      227.6      460.4 |     688.0
            |      22.42      29.70 |     27.29
------------+-----------------------+----------
   SA-Agree |        647      1,186 |     1,833
            |      606.4    1,226.6 |   1,833.0
            |      77.58      70.30 |     72.71
------------+-----------------------+----------
      Total |        834      1,687 |     2,521
            |      834.0    1,687.0 |   2,521.0
            |     100.00     100.00 |    100.00
```

All the figures are the same as what has been produced before, but now each cell contains the frequencies that would have been expected if the two variables were completely independent. For example, 647 nonattendees actually said they agree or strongly agree that they like to take risks, but based on the distribution of each variable, if the two were unrelated only 606.4 nonattendees would be expected to agree or strongly agree. In each cell, the difference between the observed and expected frequencies is substantial, further suggesting a relationship between the two variables.

Most users, however, would prefer not to have to calculate such statistics completely by hand, which is why Stata offers an option to directly produce the chi-square statistic. If you think about what you would tell a smart colleague to have him or her produce such a chi-square test, you will probably identify the correct option to invoke in Stata: `-chi-`. (The full option name actually is `-chi2-` but the acceptable, abbreviated version `-chi-` is slightly more intuitive.) As before, to produce this additional figure type the same command as you did above simply adding the `-chi-` option; `tab agrisk gender, col expected chi` into the Command line and press **Enter**. As you might surmise, the results look exactly like what was produced above, but now the following information on the chi-square test is displayed at the bottom of the table:

```
[Table Omitted]
              Pearson chi2(1) =   14.8883   Pr = 0.000
```

This new information tells you that it is using a Pearson chi-square test, the degrees of freedom involved in the test (1), and the actual chi-square statistic (14.8883). Next it displays the probability value (*p* value) of obtaining the given statistic with the current degrees of freedom. Here the *p* value is .000. Because this *p* value is less than the standard significance value of .05, you would reject the null hypothesis and conclude that there is in fact a significant relationship between college attendance and liking to take risks.

As was mentioned above, the ordering of the variables in the command line does not matter for the figures that are produced in a cross-tabulation (except for the column and row percentages). Similarly, the chi-square statistic is not affected by the placement of the variables into rows or columns. If you were to switch `agrisk` and `cu_attco` in the command line, the chi-square statistic would be exactly the same.

MEASURES OF ASSOCIATION

While the chi-square statistic is an excellent indication of whether two variables are significantly related, it does not provide an indication of the strength of that relationship. Often a great deal of importance is placed on the significance of a relationship, while the magnitude of the relationship is overlooked. Surely, knowing whether the relationship is statistically significant is important. But understanding whether that relationship is very weak or very strong is similarly important for making conclusions about what is happening in the real world.

There are several measures of association that can be used to assess the strength of a relationship between two nominal or ordinal variables. Two of the most common measures of association for nominal and ordinal variables are gamma (sometimes referred to as Goodman and Kruskall's gamma) and Kendall's Tau-*b*.[1] Stata makes producing each of these figures quite straightforward. Just as above, when you needed an additional figure it only required including a new option to the end of the `-tab-` command line. And similar to chi-square and the expected frequencies, the options to produce gamma and Kendall's Tau-*b* are intuitive: `-gamma-` and `-taub-`, respectively. Type `tab agrisk cu_attco, col expected chi2 gamma taub` into the Command window and press **Enter**. The following new figures are listed at the bottom of the table:

```
[Table Omitted]
         gamma =  -0.1875  ASE = 0.048
Kendall's tau-b =  -0.0768  ASE = 0.019
```

Both gamma and Kendall's Tau-*b* are what are called "symmetric measures of association," meaning it does not matter which variable is designated as the independent or dependent variable. As with chi-square, you can switch the ordering of the variables in the command line and these figures will not change. Additionally, both measures can vary from negative 1 to positive 1, with values at either extreme indicating a strong relationship and values near 0 indicating a weak relationship.

Gamma and Kendall's Tau-*b* are both negative in the example. The sign depends on how the two variables are coded. Having attended some college and agreeing with the statement about taking risks are both coded 1. Therefore, a negative relationship on these measures suggests that college attendees are *less* likely to agree with the statement about taking risks. The gamma value of –.1875 and Kendall's Tau-*b* value of -.0768 both suggest a moderately strong relationship between college attendance and liking to take risks. Specifically, they can be interpreted as showing that knowing whether a young adult attended college improves the prediction of his or her agreement with liking to take risks by 19% and 8%, respectively.

---

[1]Gamma and Kendall's Tau-*b* should only be used with nominal variables if they are dichotomous (e.g., male vs. female). Although there are measures of association for multicategory nominal variables, they can be a bit complex to calculate and suffer from potential reliability problems (e.g., lamda). Stata therefore does not include these figures as a standard option. It is generally advisable to rely on the percentage differences as a method for interpreting the strength of the association between two multicategory nominal variables.

ELABORATION

So far you have shown a clear relationship between college attendance and liking to take risks, such that young adults who do not attend college like to take risks more than young adults who do attend some college. A critic, however, may argue that your proposed causal relationship is flawed. Specifically, this critic might contend that gender is a spurious factor driving your observed relationship. There may not be any relationship between college attendance and liking to take risks. Rather, females might be both more likely to attend college and be averse to risk, which is why you have observed the negative relationship between college attendance and risk taking.

Testing the possible confounding influence of gender requires what is referred to as "elaboration." Using an elaboration model means taking account of a third (or additional) factor when examining the primary bivariate relationship in question. In other words, the elaboration method shows whether the main relationship (e.g., college attendance and risk taking) holds up for all groups of a third variable (e.g., males and females). If college attendance shows a relationship to risk taking for both males and females, then gender would not be a spurious factor causing the initial relationship.

Conducting the elaboration method requires that you examine the main relationship for all the categories of the third factor. In the present example, this means you need to test the relationship between college attendance and liking to take risks for just males and just females. You already have learned one way that you can have Stata produce the necessary output. You could use an `-if-` statement, using the gender variable (i.e., `-if gender==0-`), in the `-tab-` command you used above to restrict the cross-tabulation to just men. Then you would execute the same command line but alter the `-if-` statement to isolate the table to just females (i.e., `-if gender==1-`).

This combination of commands definitely would work, but Stata offers a shortcut command that can reduce these two commands into one. Thinking intuitively, you are now asking your smart colleague to construct the table *by* each category of the gender variable. Hence the command name is `-by-`. The structure of commands that use `-by-` is only slightly different.

```
by varname(s): command varname(s) [if varname==value]
[, options]
```

The second half of the command, after the ":", is exactly the same as every other command that has been covered so far. The first part of a `-by-` command is a way to tell Stata to perform the command, that follows the colon, "by" each of the categories (or combination of categories) in the variables specified in the `-by-` clause.

There is one, admittedly not intuitive, component to using the `-by-` command. Stata needs the variable that is specified in the `-by-` clause to be "sorted," meaning that the data need to be arranged with all the cases in each of the categories ordered together. There is a separate command -sort- (e.g., -sort gender-) that would perform this operation, which would then allow the -by- prefaced command to work. Again, Stata offers a shortcut. Instead of only typing -by-, Stata allows you to enter -bysort-, which automatically sorts the data by the variable(s) that follow -bysort- and runs the command across each possible category. The structure, therefore, may be more effectively thought of as

```
bysort varname(s): command varname(s) [if varname==
value] [, options]
```

A Closer Look: Using -bysort- as a Universal Tool

The -bysort- command is most clearly applicable for use with the elaboration method. Its use, however, does not stop there. For example, Chapter 4 covered how to produce descriptive statistics on the bmi variable. It may be important to look at those figures separately for males and females. Of course, two seperate -sum- commands could be used with -if- statements. To save time, type bysort gender: sum bmi in the Command window and press **Enter**. The following results are shown:

```
-> gender = Male

    Variable |       Obs        Mean    Std. Dev.       Min        Max
-------------+--------------------------------------------------------
         bmi |      1224    25.36675    5.192121   14.22837   63.49296

----------------------------------------------------------------------

-> gender = Female

    Variable |       Obs        Mean    Std. Dev.       Min        Max
-------------+--------------------------------------------------------
         bmi |      1285      24.138    5.022598   14.01495   48.81944
```

The results now clearly show that body mass index is slightly higher, on average, among young males (25.37) than females (24.13). At least one of the outlier cases (63.49) is a male, which may bias their true average.

The -bysort- command can be used with virtually any Stata command. Whenever you need to perform some operation across categories of another variable, the -bysort- command is the most effective tool.

To have the cross-tabulation between liking to take risks and college attendance elaborated by gender, type `bysort gender: tab agrisk cu_attco, col expected chi2 gamma taub` into the Command window and press **Enter**. The following results are displayed:

```
-> gender = Male

+--------------------+
| Key                |
|--------------------|
|     frequency      |
| expected frequency |
| column percentage  |
+--------------------+

  Agree or |
  Not with |
    Taking |   (cu_attendcoll_13)
 Risks (rc | Ever attended college
     P:18) |        No        Yes |     Total
-----------+----------------------+----------
  No Agree |        78        181 |       259
           |      98.4      160.6 |     259.0
           |     16.74      23.82 |     21.13
-----------+----------------------+----------
  SA-Agree |       388        579 |       967
           |     367.6      599.4 |     967.0
           |     83.26      76.18 |     78.87
-----------+----------------------+----------
     Total |       466        760 |     1,226
           |     466.0      760.0 |   1,226.0
           |    100.00     100.00 |    100.00

          Pearson chi2(1) =   8.6843   Pr = 0.003
                    gamma =  -0.2172  ASE = 0.072
          Kendall's tau-b =  -0.0842  ASE = 0.027

----------------------------------------------------------

-> gender = Female

+--------------------+
| Key                |
|--------------------|
|     frequency      |
| expected frequency |
| column percentage  |
+--------------------+
```

```
  Agree or |
  Not with |
    Taking |   (cu_attendcoll_13)
 Risks (rc | Ever attended college
     P:18) |        No        Yes |     Total
-----------+----------------------+----------
  No Agree |       109        320 |       429
           |     121.9      307.1 |     429.0
           |     29.62      34.52 |     33.13
-----------+----------------------+----------
  SA-Agree |       259        607 |       866
           |     246.1      619.9 |     866.0
           |     70.38      65.48 |     66.87
-----------+----------------------+----------
     Total |       368        927 |     1,295
           |     368.0      927.0 |   1,295.0
           |    100.00     100.00 |    100.00

         Pearson chi2(1) =    2.8555   Pr = 0.091
                   gamma =   -0.1122  ASE = 0.066
         Kendall's tau-b =   -0.0470  ASE = 0.027
```

The top table, as the header denotes, contains only males, while the bottom table contains only females. If gender was a spurious factor causing the relationship between college attendance and liking to take risks, then the original bivariate relationship (i.e., college attendance being negatively related to taking risks) should disappear or greatly decrease when the relationship is separated by gender. Therefore, the presence, significance, and strength of the relationship between college attendance and taking risks for males only and females only should be compared with the original bivariate relationship.

Starting with females, there is some evidence of a spurious relationship. The percentage difference between attendees and nonattendees who agree or strongly agree that they like taking risks has decreased from 7% to just less than 5%. Similarly, both the gamma and Kendall's Tau-*b* are weaker (i.e., closer to 0) for the female-only table. And the chi-square (.091) is no longer less than the standard a level of .05, meaning you would fail to reject the null hypothesis and conclude that there is no relationship between college attendance and liking to take risks among young adult females.

The table for males shows a different story. The percentage difference in agreeing with liking to take risks is similarly 7%, as it was in the full sample table. The gamma and Kendall's Tau-*b* have actually increased slightly, suggesting the relationship is minimally stronger for males. Finally, the chi-square (.03) is less than the standard a of .05, meaning you would fail to reject the null hypothesis and conclude that there is a relationship between college attendance and risk taking for young adult males.

This pattern of results provides some evidence for what is called a "conditional relationship." The relationship between college attendance and liking to take risks is conditional on young adults being male. In other words, attending college does reduce young adults liking to take risks but only for males. For females, there is no relationship between going to college and liking to take risks.

MULTIVARIATE BAR GRAPHS

The information that was presented in this elaborated cross-tabulation can be displayed in graphical form. The best type of graph to show this information is a bar graph. Before starting to construct the graph, it may be helpful to think about what exactly you want to show. For this example, the graph would need to display the percentage of respondents that agree or strongly agree with the taking risks statement by college attendance, and then each of those comparisons would need to be made for males only and females only. As you use the point-and-click menus, keeping this overall plan in mind helps identify which variables should be placed in each box.

Start by clicking on the **Graphics** menu and then the **Bar Chart** option. You see a window like the one below:

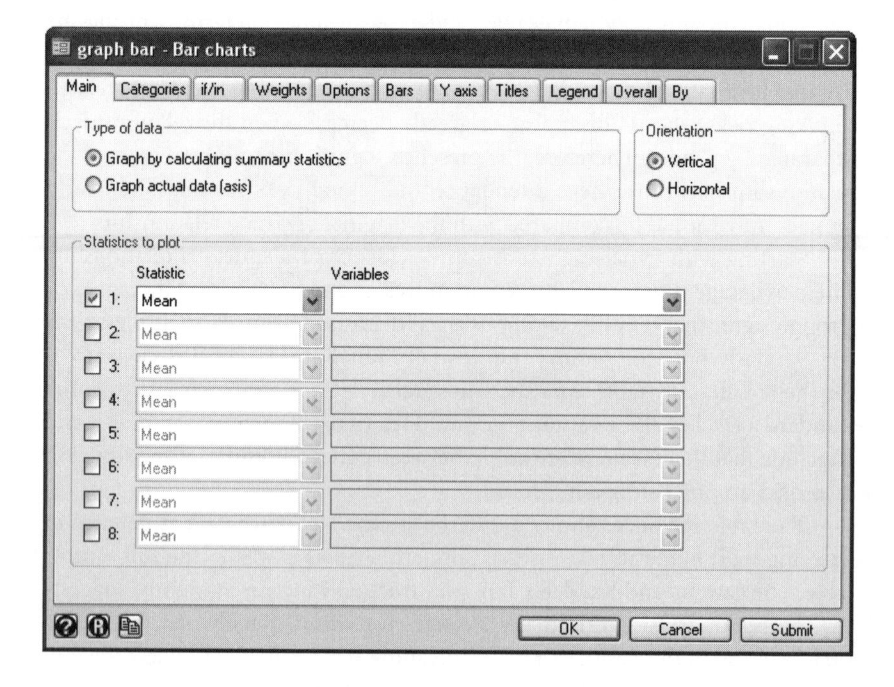

This main page of the bar charts menu is where you place the primary variable you want to be displayed. If you look above, the main goal is to show the percentage of respondents who strongly agree or agree with liking to take risks. Therefore, you should type (or use the drop-down menu to find) `agrisks` in the **Variables** box. The **Statistic** drop-down menu to the left of this is where you specify which figure about the variable is displayed. "Percentage" would seem to be the clear option, but when a variable is dichotomous and is coded 0/1, as the `agrisk` variable is, the **Mean** will produce the same figure. So that box does not need to be altered.

Next, select the **Categories** tab to display the following window:

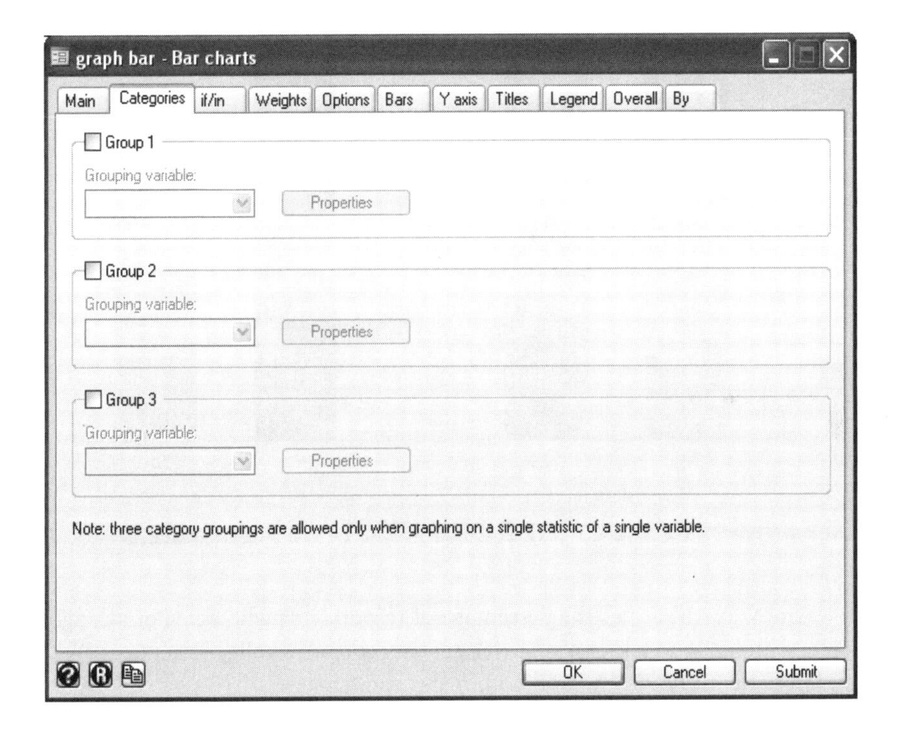

The **Grouping variable** boxes are where you specify by what categories the main statistic should be displayed. The grouping variables are processed in order. Looking back at the statement of what you are trying to present, you want the percentage of agreement with liking to take risks by college attendance and then by gender. Thus, you should select the radio button next to **Group 1** and then type (or use the drop down menu to find) `cu_attco` in the **Grouping variable** box. Then do the same but with `gender` with the

Group 2 – Grouping variable box. Then select **OK**. When you do this, the following window will appear:

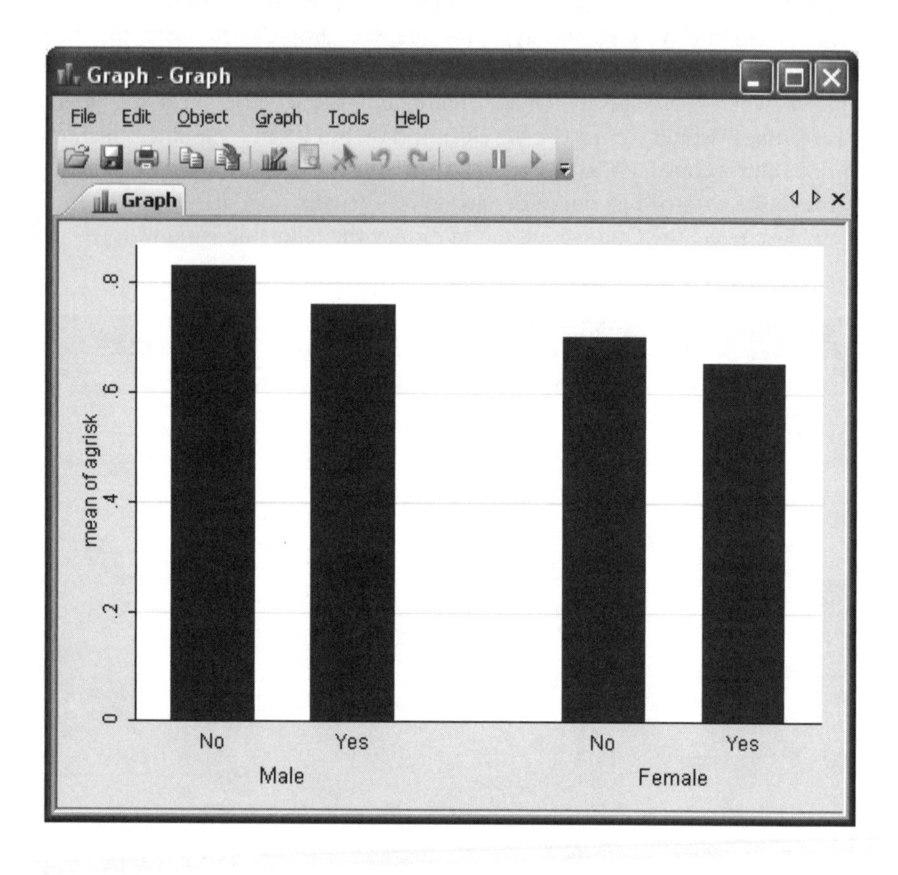

The Y-axis indicates the percentage of respondents who agree or strongly agree with liking to take risks. Then each bar represents college attendees (Yes) and nonattendees (No), within both males and females.

Although this graph displays the same information as the elaborated cross-tabulation, the visual depiction more clearly illustrates the overall pattern. First, it shows that a higher percentage of males like to take risks than females. In fact, a higher percentage of male college attendees agree that they like to take risks than female *nonattendees*. Similarly, the difference in agreement between the two attendance groups is clearly larger for the males than for the females.

One aspect of the graph that could be improved is the labeling of attendee versus nonattendee bars. The "Yes" and "No" value label attached to the actual variable makes sense because the variable label clearly states that the variable refers to whether the person has ever attended college. Without that variable label, the value labels are not as informative. To change these labels within the graph only, return to the Categories window and click on the Properties box next to where you entered `cu_attco`. The window shown below appears:

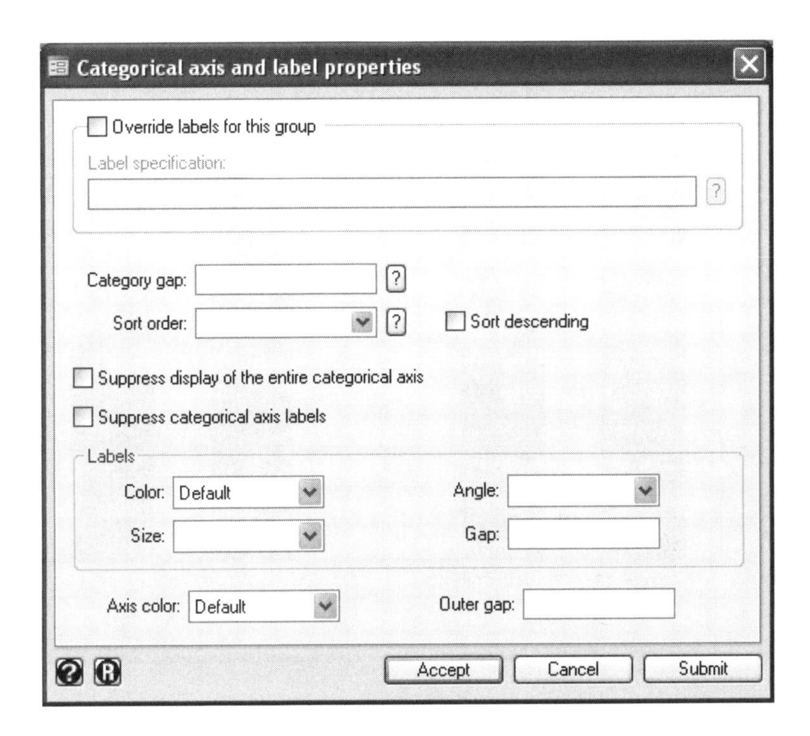

In this window, select the radio button next to Override labels for this group. In the Label specification box, you can type what you want the new labels to be. First you must type the category to which the label within the graph should be applied. Even though the noncollege attendees are coded as 0 on the variable, they technically represent the first group and the college attendees are the second group. The correct specification for this window is: `1 "No College" 2 "Attended College"`. [Note that this method for assigning labels is different from the `-lab def-` command and is particular to using the point-and-click graph label specification. Clicking on the ? icon reminds users of this difference.] After you have filled in the new label specification, click Accept and then OK. The new graph will look like the one shown below:

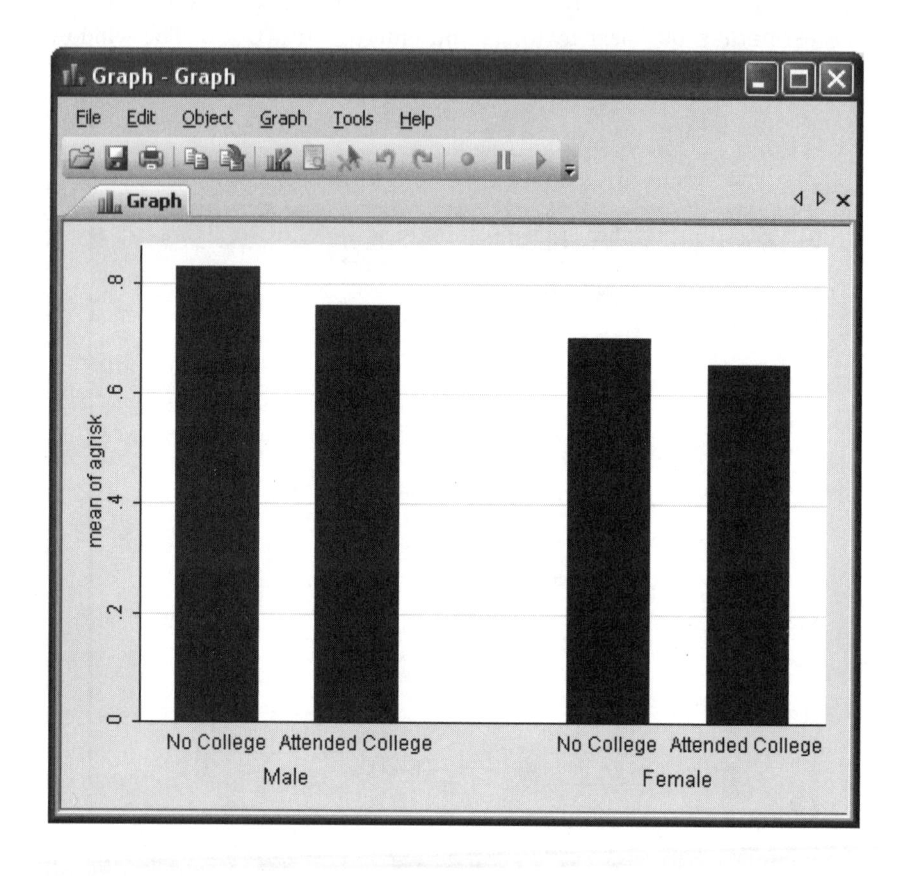

Summary of Commands Used in This Chapter

```
*Cross Tabulations
tab1 risks cu_attco
tab risks, nol
recode risks (1/2=1) (3/5=0), gen(agrisk)
tab risks agrisk
lab def ynagree 0 "No Agree" 1 "SA-Agree"
lab val agrisk ynagree
lab var agrisk "Agree or Not with Taking Risks (rc
P:18)"
tab agrisk
```

```
tab agrisk cu_attco

tab agrisk cu_attco, col
tab agrisk cu_attco, col nokey nofreq

*Chi-Square
tab agrisk cu_attco, col expected
tab agrisk cu_attco, col expected chi

*Measures of Association
tab agrisk cu_attco, col expected chi gamma taub

*Elaboration
tab agrisk cu_attco if gender==0,
col expected chi2 gamma taub


bysort gender: tab agrisk cu_attco, col expected chi2
gamma taub

*A Closer Look: Recoding for Direction
recode risks (1=5) (2=4) (3=3) (4=2) (5=1), gen(likerisk)
tab likerisk

*A Closer Look: Using -by- as a Universal Tool
bysort gender: sum bmi
```

Exercises

Use the original `Chapter 5 Data.dta` for the following problems. [Optional: Complete the exercises using a do file and save the results using a log file. See Chapter 3 for an explanation of how to use these files.]

1. Use a cross-tabulation to examine the prediction that young adults who report religion being more important (`faith1`) will care more about the elderly (`crelder`).
2. Produce a chi-square statistic to further investigate whether the importance of religion is significantly related to caring for the elderly.
3. Use the gamma and Kendall's Tau-*b* statistics to determine the strength of the relationship between importance of religion and caring for the elderly.

4. Use the elaboration method to investigate whether the relationship between religious importance and caring for the elderly is actually spuriously caused by attendance at religious services.

    4a. Create a dichotomous variable that indicates whether the respondent attends religious services (`attend`) many times a year or more.

    4b. Produce two cross-tabulations of religious importance and caring for the elderly: one for respondents who do not attend religious services frequently and one for respondents who do attend religious services frequently.

6

Relationships Between Different Measurement Levels

Thus far all the statistical techniques that have been covered involved variables measured at a nominal or ordinal level. Research projects frequently pose questions about the relationship between variables that have numerous categories, sometimes one for each respondent in a sample. These types of variables are typically referred to as interval-ratio variables. Such variables are also generally based in some real measurement unit, meaning the numeric value attached to these variables has some real-world meaning. Examples of interval-ratio variables are income, IQ scores, age, and cholesterol levels.

Commonly, a researcher is interested in how an ordinal or nominal variable is related to an interval-ratio variable. For example, you might wonder whether the religious denomination of young adults affects how many times they volunteer. Amount of volunteering would most likely be an interval-ratio variable, whereas denomination affiliation would be a nominal variable. This chapter examines two strategies that can be employed to answer this type of question.

All the examples that follow use the `Chapter 6 Data.dta`, available at **www.sagepub.com/longest**. This data set includes a more extensive set of variables from the National Study of Youth and Religion (NSYR) data and contains the full Wave 3 sample of 2,532 young adults. All the missing cases have been replaced with appropriate missing codes in this data set, with `.d` referring to a response of "Don't Know," `.r` to a response of "Refused," and `.s` to cases that were legitimately skipped out of a question based on the survey design (i.e., skip pattern).

Testing Means

The research question stated above was whether religious denomination affiliation is related to the amount of volunteering among young adults. Another way to pose this question would be "Is the average number of times volunteered the same for all religious denominations." Wording the question in this way highlights the fact that analytic strategies that look at the relationship between an interval-ratio and nominal variable are primarily focused on comparing the mean level of the interval-ratio variable across each group of the nominal variable.

The NSYR data contain a variable, `volnum2`, that stems from a question asking "About how many times in the last year did you do volunteer work or community service work?" The respondents could answer any whole number between 1 and 80. This survey item was preceded by a question that asked respondents if they had done any volunteer work. Anyone who reported that they had not done any volunteer work in the past year was skipped out of the question about how many times they had volunteered. Therefore, this variable can be recoded to change respondents who are set as the missing code for skip (`.s`) to 0 because their answer to the previous question clearly shows they did not do any volunteer work.

To make this change, type `recode volnum2 (.s=0), gen(freqvol)` in the Command window and press **Enter** (see The 5 Essential Commands: recode section of Chapter 2 for more information on the `-recode-` command). Then produce a detailed summary statistics report of the newly generated, interval-ratio variable by typing `sum freqvol, det` into the Command window and press **Enter**.

```
            RECODE of volnum2 ((volnum2_w3) [IF HAS
           VOLUNTEERED] H:13. About how many times
-------------------------------------------------------------
      Percentiles      Smallest
 1%            0              0
 5%            0              0
10%            0              0       Obs                2525
25%            0              0       Sum of Wgt.        2525

50%            0                      Mean           5.245941
                        Largest       Std. Dev.      13.06794
75%            4             80
90%           12             80       Variance       170.7711
95%           30             80       Skewness       3.970169
99%           80             80       Kurtosis       19.91374
```

The table first shows that the -recode- command saved most of the missing cases as the number of valid observations for this variable is 2,525 (out of the 2,532 possible respondents). Second, it indicates that, on average, young adults volunteered just over five times in the last year. The median for this measure is 0, meaning that more than half of the respondents reported not volunteering at all in the last year. This relatively high amount of 0 cases also skews the distribution, which can be seen by the mean (5.25) being greater than the median. The distribution is also relatively spread out, shown by a standard deviation of 13.07.

CONFIDENCE INTERVALS

As noted above, the primary methods for assessing relationships between an interval-ratio variable and a nominal variable involve tests of the mean. Before examining its relationship with religious denomination affiliation, it is helpful to examine more closely the mean of the volunteering variable.

A way to gain more information about the mean of a variable is to construct its confidence interval. The mean of a sample is based, in part, on the specific sample that was taken to produce the variable. Each sample might be slightly different, which could produce slightly different means for the same variable. A confidence interval is a way to adjust for these minor differences and provide a range in which the true average of the population should fall.

The command to produce a confidence interval in Stata is -ci-. Type ci freqvol into the Command window and press **Enter** to present the following results:

```
    Variable |    Obs        Mean    Std. Err.   [95% Conf. Interval]
-------------+---------------------------------------------------------
     freqvol |   2525    5.245941    .2600618    4.735984    5.755897
```

The output looks very similar to that produced by the brief -sum- command. But the -ci- command displays the standard error of the mean (Std. Err.), instead of the standard deviation, and the 95% confidence interval of the mean, instead of the minimum and maximum values. For the frequency of volunteering, the 95% confidence interval is 4.74 and 5.76. Or in other words, you can be 95% confident that the true population average of young adults' frequency of volunteering is between 4.74 and 5.76.

The confidence level displayed can be controlled by invoking the `-level(#)-` option. If you wanted to be more confident about the true population mean, you could type `ci freqvol, level(99)` into the Command window and press **Enter**.

```
    Variable |        Obs        Mean    Std. Err.       [99% Conf. Interval]
-------------+---------------------------------------------------------------
     freqvol |       2525    5.245941    .2600618        4.575559    5.916322
```

The table is very similar to the one shown above, except now the 99% confidence interval is displayed. As would be expected, increasing the confidence level widens the interval (i.e., to be more confident you have to give yourself a broader range to catch the mean). Now you can be 99% confident that the true population mean of number of times volunteered is between 4.58 and 5.92.

TESTING A SPECIFIC VALUE (ONE-SAMPLE *t* TEST)

In addition to producing confidence intervals, you can conduct a significance test of whether the mean of a variable is equal to a particular value in the true population. For instance, the mean for the `freqvol` variable in the NSYR data is 5.25. Imagine you heard a report on the news that claimed young adults in the United States volunteered, on average, 3 times in the past year. Even though the NSYR sample mean of volunteering in the past year is greater than 3, this single figure does not answer the question of whether the true young adult population mean is greater than 3, due to potential sampling error. To answer the latter question, you need to conduct a statistical test to see if the `freqvol` variable's mean is equal to, less than, or greater than 3.

The command to execute this test is not immediately intuitive, as you might be considering something involving "mean test." The test used to make the determination about a variable's mean is technically referred to as a *t* test because it depends on the *t* distribution. Knowing this piece of information, the command is much more intuitive: `-ttest-`.

To conduct the test for the given question, type `ttest freqvol==3` into the Command window and press **Enter**. A double equal sign is used because you are asking Stata to evaluate whether one value equals another (just as you have done when using `-if-` statements). Also, notice that you do not have to specify whether you are interested in the mean of the specified variable being greater than or less than the value. Using the double equal sign in the command line produces all three pertinent tests.

Once you have executed the command, the following results are shown:

```
One-sample t test
------------------------------------------------------------------------------
Variable |    Obs        Mean    Std. Err.   Std. Dev.   [95% Conf. Interval]
---------+--------------------------------------------------------------------
 freqvol |   2525    5.245941    .2600618    13.06794    4.735984    5.755897
------------------------------------------------------------------------------
    mean = mean(freqvol)                                          t =   8.6362
Ho: mean = 3                                     degrees of freedom =     2524

    Ha: mean < 3                 Ha: mean != 3                 Ha: mean > 3
 Pr(T < t) = 1.0000         Pr(|T| > |t|) = 0.0000          Pr(T > t) = 0.0000
```

The top half of the results replicates the information provided by the default `-ci-` command. Just below this table, the pieces involved in the test are summarized. The left-hand side shows that the mean of `freqvol` is being tested followed by the null hypotheses being set as the mean being equal to 3. The right-hand side presents the calculated *t* statistic (8.6362) and the degrees of freedom (2,524) used in the test.

Across the bottom portion of the output, three separate results are shown. Each one presents a different alternative hypothesis. The middle test is simply whether the mean is equal to 3 or not. The *p* value (`Pr(|T| > |t|)` of .0000 is less than the standard a level of .05, meaning you can reject the null hypothesis and conclude that the mean number of times volunteered in the population of young adults is significantly different from 3. The results on the left side present the results for the test of whether the mean is less than 3, and the ones on the right are for the test of whether the mean is greater than 3.

The results on the right should be used for the hypothesis claiming young adults volunteer more than 3 times in a year. The *p* value (`Pr(|T| > |t|)` for this test is also .0000. Because this value is less than the standard a level of .05, you can reject the null hypothesis and conclude that the mean number of times volunteered in the population of young adults is significantly greater than 3.

TESTING THE MEAN OF TWO GROUPS (INDEPENDENT-SAMPLES *t* TEST)

Now that you have thoroughly examined the mean of the `freqvol` variable for the entire sample, you can proceed to examine whether the average number of times volunteered differs by religious denomination. Testing the means of two groups is referred to as an independent-samples or two-sample test. The wording can be confusing because it makes it seem as though two unique samples (i.e., data sets) have to be used. Although two different samples can be used with these tests, two groups from within one data set can also be used. In the latter sense, you might think as though you are taking a sample of all the Catholics from the NSYR data, for example, and comparing them with a sample of all the non-Catholics in the NSYR data set.

The NSYR contains numerous questions and variables that could be used to conduct this test. The ultimate research question is whether religious denomination affiliation is related to the number of times young adults volunteer in a given year. Before examining the differences across specific religious denominations, it would be reasonable to see if there is a difference in volunteering between young adults who identify with any religious denomination and young adults who do not report being affiliated with any denomination.

The `bntranr` variable is a dichotomous indicator of respondents who are classified as being Not Religious. This variable's name is not especially clear, which makes using the `-rename-` command an effective strategy, although not completely necessary.

Type `rename bntranr notrel` into the Command window and press **Enter**. You will see that the name of the variable changes in the Variables window. Now produce a frequency distribution of this variable by typing `tab notrel` into the Command window and pressing **Enter**.

```
(bntranr_w3 |
) Dummy for |
        Not |
  Religious |      Freq.     Percent        Cum.
------------+-----------------------------------
          0 |      1,910       75.43       75.43
          1 |        622       24.57      100.00
------------+-----------------------------------
      Total |      2,532      100.00
```

The table shows that almost 25% of young adults are categorized as being Not Religious (coded 0).

The command to conduct a *t* test of the mean of the `freqvol` variable over the two categories of this `notrel` variable is similar to the one used to conduct a *t* test of a specific value. Enter the command and then the variable: `ttest freqvol`. Now you need to tell Stata to compare the means of the variable being tested by the two categories of a second variable. To do so, you invoke the `-by(varname)-` option.

Type `ttest freqvol, by(notrel)` into the Command window and press **Enter**. The following results are displayed:

```
. ttest freqvol, by(notrel)

Two-sample t test with equal variances
------------------------------------------------------------------------------
   Group |     Obs        Mean    Std. Err.   Std. Dev.   [95% Conf. Interval]
---------+--------------------------------------------------------------------
       0 |    1903    5.792958     .311302    13.58005    5.182429    6.403488
       1 |     622    3.572347    .4492347    11.20388    2.690144    4.454551
---------+--------------------------------------------------------------------
combined |    2525    5.245941    .2600618    13.06794    4.735984    5.755897
---------+--------------------------------------------------------------------
    diff |            2.220611    .6020626                1.040024    3.401199
------------------------------------------------------------------------------
    diff = mean(0) - mean(1)                                      t =   3.6883
Ho: diff = 0                                     degrees of freedom =     2523

    Ha: diff < 0                 Ha: diff != 0                 Ha: diff > 0
 Pr(T < t) = 0.9999         Pr(|T| > |t|) = 0.0002          Pr(T > t) = 0.0001
```

The results, especially the bottom portion, are similar to those produced by the *t* test of a specific value. The top half of the table is slightly different. Now the means, standard errors, standard deviations, and 95% confidence intervals for the `freqvol` variable are presented for each group of the `notrel` variable. The results show that young adults classified as Not Religious volunteered, on average, 3.57 times in the last year, compared with 5.79 times, on average, among young adults who claim some religious identification. The next row presents the same result for the two groups `combined` (i.e., for the total sample), and the final row of the upper tables presents the statistics for the difference between the two groups. The difference in the mean number of times volunteered in the last year for Not Religious and religious young adults is 2.22.

The bottom portion of the table presents the results of the significance test of whether the difference in means (2.22) is different, less than or greater than 0. In other words, it tests if the difference between the two groups' means is significant. The upper right shows the calculated *t* statistic (3.6883) and the degrees of freedom used in the test (2,523).

As with the previous `-ttest-` output, three separate results are presented at the bottom of the display. The middle portion reports the results from the test of whether the difference in means is significantly different from 0. The *p* value (`Pr(|T| > |t|)`) of .0002 is less than the standard a level of .05, which means you can reject the null hypothesis and conclude that the mean number of times volunteered is different for Not Religious and religious young adults. Similalry, the right side of these results also shows a *p* value (`Pr(|T| > |t|)`) that is less than .05. These results indicate that you can reject the null hypothesis and conclude that the average number of times volunteered in the last year is greater among religious young adults than among young adults who are Not Religious.

Analysis of Variance (ANOVA)

The *t* test of means is an excellent way to assess the relationship between a dichotomous nominal variable and an interval-ratio variable. The original research question, however, asked whether volunteering varied by specific religious denomination. Even an extremely condensed classification of denomination would have to include more than two categories, making a *t*-test comparison of means infeasible.

The appropriate strategy for this type of research question is an analysis of variance, typically referred to as ANOVA. An ANOVA analysis is somewhat similar to the comparison of the means of the two groups performed above, but it compares the means of multiple groups simultaneously.

One of the more condensed variables that captures denomination affiliation in the NSYR data is `i_religi`. As above, this variable name is somewhat confusing, so use the `-rename-` command to change it to `denom` (i.e., `-rename i_religi denom-`). Next, produce a frequency distribution of this nominal variable to see all of the categories it contains by typing `tab denom` into the Command window and pressing **Enter**.

```
(tradrel_w3) Identical |
 to relatt_w3 but uses |
 identification info on |
          non-attenders |      Freq.     Percent        Cum.
------------------------+-----------------------------------
Evangelical Protestant |        714       28.20       28.20
   Mainline Protestant |        259       10.23       38.43
      Black Protestant |        189        7.46       45.89
              Catholic |        443       17.50       63.39
          Not religious |        622       24.57       87.95
         Other religion |        305       12.05      100.00
------------------------+-----------------------------------
                  Total |      2,532      100.00
```

The six denomination groups are somewhat equally distributed, with Evangelical Protestant having a slightly greater percentage of the respondents than the rest, and Black Protestant being somewhat smaller.

To get a general sense of the difference in the level of volunteering across these six denominations, it may be helpful to produce a table of the means for each group. There are several methods that would produce these results. For example, you could execute six different `-sum freqvol if denom==#-` commands, filling in # for each category's code, but this approach would be quite time-consuming. A quicker option would be to use the `-bysort denom:-` prefix to the `-sum freqvol-`, as shown in Chapter 5 (A Closer Look: Using `-bysort-` as a Universal Tool). This method may be faster, but the means would be spread out in the display and slightly difficult to compare. The most effective method would be to use the `-tabstat-` command and invoke its `-by(varname)-` option. The `-by(varname)-` option with `-tabstat-` produces all the requested statistics for each category specified by varname.

Remember any of these strategies would have produced similar results, and in the end that is all that matters. Do not become frustrated by thinking you must know and remember every possible command to produce a given outcome. As long as the output produces what the research project needs, you have used the "correct" command.

Type `tabstat freqvol, by(denom)` into the Command window and press **Enter**. The following table is displayed:

```
Summary for variables: freqvol
     by categories of: denom ((tradrel_w3) Identical to relatt_w3
but uses identification info on non-attenders)

               denom |      mean
---------------------+----------
    Evangelical Prot |  5.115331
    Mainline Protest |    6.6139
    Black Protestant |  4.074074
            Catholic |  5.800454
        Not religious |  3.572347
       Other religion |  7.742574
---------------------+----------
               Total |  5.245941
--------------------------------
```

The table shows the average number of times volunteered in the last year across the different denomination groups. The most frequent volunteers, at almost 8 times a year, are young adults in an Other religion, followed by Mainline Protestants, who volunteer just over 6.5 times a year. The least frequent volunteers are young adults who are Not Religious (3.57). The table shows that there is a difference in the average amount of times young adults volunteer based on their religious denomination.

The command to conduct the statistical ANOVA test of the significance of these differences, as you might intuitively guess, is `-anova-`. Because ANOVA is not a symmetrical test, the ordering of the variables entered after the command is important. The dependent variable, here frequency of volunteering, should be entered first, directly after the command, followed by the independent variable(s).[1]

Type `anova freqvol denom` into the Command window and press **Enter.**

```
. anova freqvol denom

                  Number of obs =     2525     R-squared     =  0.0105
                  Root MSE      = 13.0121      Adj R-squared =  0.0085

      Source |  Partial SS    df       MS           F     Prob > F
  -----------+----------------------------------------------------
       Model |     4522.77     5  904.553999       5.34     0.0001
             |
       denom |     4522.77     5  904.553999       5.34     0.0001
             |
    Residual |  426503.501  2519  169.314609
  -----------+----------------------------------------------------
       Total |  431026.271  2524  170.771106
```

There are numerous results presented, but the key figures are the ones listed in the `Partial SS` column and the `Prob > F` column. The number in the `denom` row of the `Partial SS` column (4522.77) shows the degree of variation in frequency of volunteering that occurs across groups of religious denomination. In the next row, titled `Residual`, the figure in the `Partial SS` column (426503.501) lists the amount of variation in frequency of volunteering within the individual denomination groups.

The *p* value associated with the variation across groups, listed in the `denom` row, indicates whether there is a significant difference in frequency of volunteering by denomination. The presented value of .0001 is below the standard a level of .05, meaning you would reject the null hypothesis and conclude that the frequency of volunteering of young adults is significantly different based on religious denomination affiliation.

Summary of Commands Used in This Chapter

```
*Testing Means
recode volnum2 (.s=0), gen(freqvol)
```

[1]The dependent variable in an ANOVA analysis should be measured at the interval-ratio level. The first variable typed after the `-anova-` command, therefore, always should be an interval-ratio level variable.

```
sum freqvol, det

*Confidence Intervals
ci freqvol
ci freqvol, level(99)

*Testing a Specific Value
ttest freqvol==3

*Testing the Mean of Two Groups
rename bntranr notrel
tab notrel
ttest freqvol, by(notrel)

*Anova
rename i_religi denom
tab denom
tabstat freqvol, by(denom)
anova freqvol denom
```

Exercises

Use the original `Chapter 6 Data.dta` for the following problems. [Optional: Complete the exercises by using a do file and save the results using a log file. See Chapter 3 for an explanation of how to use these files.]

1. Produce the confidence interval of the mean number of days of young adults' longest relationship (`longstr`).
2. Calculate the 99% confidence interval of the `longstr` variable.
3. Test the hypothesis that the average longest relationship of young adults is 1 year (365 days).
4. Examine whether the longest relationship young adults have experienced is significantly different for those who have ever cohabitated and those who have never cohabitated (`cu_cohab`).
5. Use an ANOVA analysis to investigate whether there are significant differences of young adults' longest relationship by their employment status (`employst`).

7

Relationships Between Interval-Ratio Variables

While the techniques discussed in Chapter 6 are effective when the independent variable is measured at the nominal or ordinal level, they do not apply when both the dependent and independent variables are interval-ratio variables. Numerous research questions involve these types of relationships. For example, you might examine how years of education attained influences a person's yearly income or how the average income of neighborhoods influence the number of people who vote in that neighborhood. This chapter explores several techniques for examining the relationship between these types of variables.

All the examples that follow use the `Chapter 7 Data.dta`, available at **www.sagepub.com/longest**. This data set includes the full National Study of Youth and Religion (NSYR) Wave 3 sample of 2,532 young adults. All the missing cases have been replaced with appropriate missing codes in this data set, with `.d` referring to a response of "Don't Know," `.r` to a response of "Refused," and `.s` to a case that was legitimately skipped out of a question based on the survey design (i.e., skip pattern).

Correlation

Typically, research questions concerning interval-ratio variables ideally try to assess if the independent variable causes the dependent variable. More tentatively, these studies investigate whether the independent variable is associated with the dependent variable. "Associate" in this context essentially means whether the values of the independent are systematically related to the values on the dependent variable.

For example, a researcher might be interested in determining the factors that influence the amount of hours young adults work in paid employment. One variable that may affect young adults' time spent in paid employment is the number of extracurricular activities in which they are involved. In this example, hours of work is the dependent variable and number of extracurricular activities is the independent variable. You might hypothesize that as the number of extracurricular activities young adults participate in increases the amount of hours they work for pay should decrease (i.e., a negative relationship) because they simply do not have time for both. Or you could reasonably hypothesize that as the number of extracurricular activities increases the number of work hours might also increase (i.e., a positive relationship) because people who are involved in one arena tend to be highly involved in all arenas.

The NSYR data contain variables that measure both these social phenomena. The first, `workhrs1`, comes from a question, "How many hours in a typical week are you currently working for pay?" Respondents could report any number from 0 to 100. The assessment of extracurricular activities was actually based on two questions. Both questions asked "How many organized activities such as groups, clubs, sports, or extracurricular activities are you involved in?" The difference is that one question asked about activities sponsored by religious organizations and the other only inquired about those not sponsored by religious organizations, resulting in two variables `relacts` and `notrelac`.

Again this situation is typical in quantitative research when using secondary data. The exact variable you need may not be ready-made. In this case, you need to generate a variable representing the total number of activities. To do so, the number of activities sponsored by religious organizations needs to be added to the number of activities not sponsored by religious organizations.

Think about how you would ask your smart colleague to accomplish this task, and your verbal request should lead you to the most effective command, which was discussed in Chapter 2. You might ask, "Please generate a new variable, called `totacts`, that is equal to the values of the `relacts` variable plus the `notrelac` variable." Replacing the necessary components with the appropriate Stata syntax produces the correct command.

Type `gen totacts=relacts+notrelac` into the Command window and press **Enter**.

Before jumping into the analysis of the relationship between these two variables, it is helpful to look at the descriptive statistics for both. Because they are interval-ratio variables, using measures of central tendency and variability is appropriate. Type `sum workhr totacts, det` into the Command window and press **Enter**. [Note that the full name of the work hours variable is `workhrs1`, but remember that variable names can be abbreviated as long as the shortened version does not overlap with another variable name in the data set. The work hours variable could be entered just as `work`, but `workhr` should prevent confusion as to what the variable means.]

```
          (workhrs1_w3 ) H:6. How many hours in a typical
                   week are you currently working for pay?
-------------------------------------------------------------
      Percentiles      Smallest
 1%            0              0
 5%            0              0
10%            0              0       Obs                2527
25%            0              0       Sum of Wgt.        2527

50%           20                      Mean            20.5002
                        Largest       Std. Dev.      19.02217
75%           40             90
90%           45            100       Variance       361.8431
95%           50            100       Skewness       .5235251
99%           70            100       Kurtosis       2.534666

                          totacts
-------------------------------------------------------------
      Percentiles      Smallest
 1%            0              0
 5%            0              0
10%            0              0       Obs                2516
25%            0              0       Sum of Wgt.        2516

50%            1                      Mean            1.54372
                        Largest       Std. Dev.      2.233683
75%            2             14
90%            4             19       Variance        4.98934
95%            5             25       Skewness       6.370865
99%            8             52       Kurtosis       114.0386
```

The results show that young adults work 20 hours per week on average, and participate in between 1 and 2 activities (1.54). Based on the standard deviations, the distributions of both measures are reasonably spread. This distribution could stem from the fact that the 25th percentile for both variables is 0, indicating that at least 25% of young adults report not working any hours and 25% report not being involved in any activities. Finally, both have more than 2,500 valid cases, so missing data does not appear to be a major problem.

SCATTERPLOTS

A very effective initial step in examining the relationship between two interval-ratio variables is to investigate the relationship visually. To do so, you need to plot each case's position on a graph based on his or her value on the two variables. This type of graph is referred to as a scatterplot or scatter diagram. In a scatterplot, one variable's values provide the scale for the X-axis (usually the independent variable) and the others are used for the Y-axis (usually the dependent variable).

Producing a scatterplot in Stata is one graph that is actually easier to execute through the Command window interface and is a very intuitive command. Your first guess is probably correct: `-scatter-`. When using the `-scatter-` command, you typically list two variables after the command. The two variables can be entered in any order, but whichever variable is listed first (directly after the command) will serve as the Y-axis, while the second variable will be plotted on the X-axis. For this reason, it is good practice to always list the dependent variable first, followed by the independent variable.

Type `scatter workhr totacts` into the Command window and press Enter. The following graph is displayed:

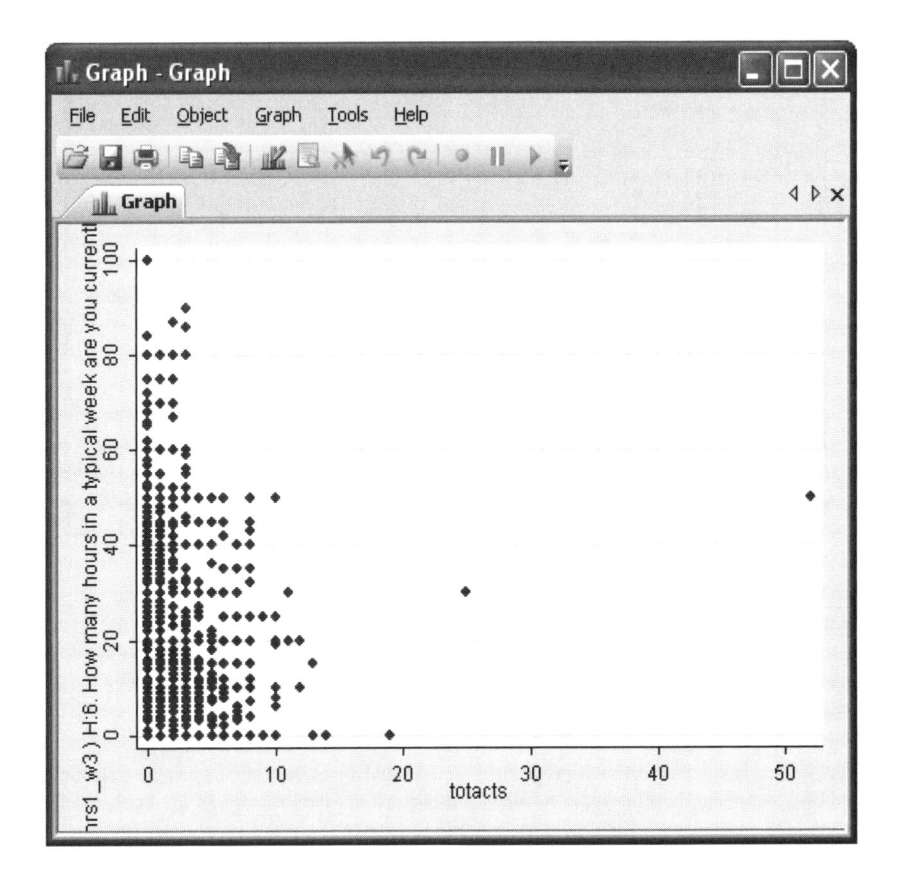

The graph shows each case as a point on the graph.[1] The most illustrative case of how a scatterplot is constructed is the point to the far right of the graph. This case participates in more than 50 activities and works around 50 hours a week.[2] Due to this extreme case, the relationship between the two variables is slightly difficult to see. To adjust the graph and gain a better depiction of the majority of cases, you can use an `-if-` statement to restrict the graph to only plot cases with more normal values on the `totacts` variable.

Type `scatter workhr totacts if totacts<20` into the Command window and press **Enter**.

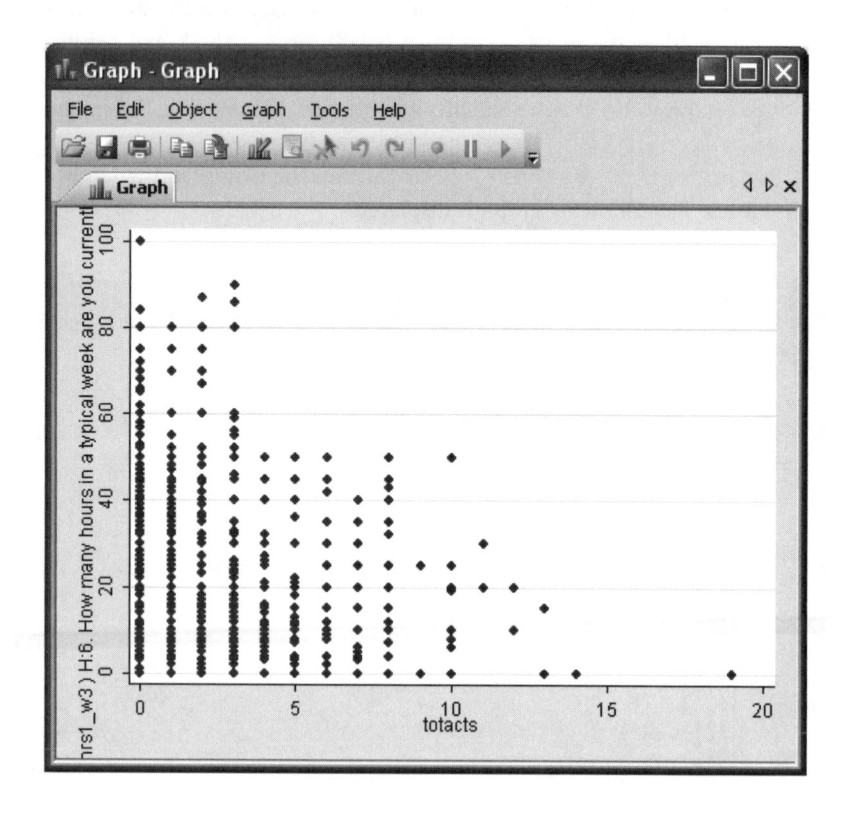

---

[1]Technically, each case is not shown because many cases have similar values on both `workhrs` and `totacts`, meaning some points represent several cases. It is possible to alter the display so that points representing more cases are clearly demarcated. Invoking the `-jitter(#)-` option after the `-scatter-` command adds volume to points that multiple cases satisfy. Increasing the # in the option increases the relative size of the volume added.

[2]In the course of a full research project, this case should be investigated as being a potentially influential outlier. The strategy for handling such cases is beyond the scope of this book, but the `-recode-` command would be a viable option for altering the value of this case (perhaps to missing or a lower, capped number of activities).

The restricted graph provides a much clearer picture of the relationship between the two variables. Most of the points that are higher on the Y-axis (i.e., work hours), are at the low end of the X-axis (i.e., number of activities), and vice versa. This pattern suggests a negative relationship, such that when young adults participate in more extracurricular activities their hours of paid employment should decrease. Or, as the number of extracurricular activities of young adults decrease, their time spent in paid employment should increase. Both statements express the same relationship, but depending on your perspective, one may be easier to understand than the other.

Now that you have established the form of the relationship (i.e., negative), the next question is how strong the relationship is. One of the first techniques that can be used to assess the strength of the relationship between two interval-ratio variables is a correlation, often referred to as Pearson's correlation coefficient or *r*. A correlation coefficient can range from negative 1 to positive 1, with values at either extreme indicating a strong relationship and values closer to 0 suggesting a weak relationship. The sign of the correlation coefficient denotes the direction of the relationship (i.e., positive or negative).

The command to produce the correlation coefficient, as you may have guessed if you were thinking intuitively, is `-correlate-` (shortened `-corr-`). A correlation coefficient is a symmetrical measure, meaning the ordering of the variables after the command does not matter. As noted above, however, it can be good practice to get into the habit of listing the dependent variable first (after the command) followed by the independent variable.

Type `corr workhr totacts` into the Command window and press **Enter**. The following table is shown:

```
(obs=2515)

             | workhrs1  totacts
-------------+------------------
    workhrs1 |   1.0000
     totacts |  -0.1525   1.0000
```

The table lists the number of cases (2,515) that were used in calculating the correlation coefficient. The actual correlation coefficients are displayed as a matrix such that each coefficient represents the correlation of the two variables intersecting in that particular cell. For example, the first figure, 1.0000, is the correlation of `workhrs1` with `workhrs1`. Because this is the correlation for the variable with itself, it will always be a perfect, positive correlation.

The correlation coefficient that you are interested in for this example is between `workhrs1` and `totacts`. This coefficient, -.1525, is listed in the cell for the `workhrs1` column and `totacts` row. As the scatterplot suggested, the correlation coefficient indicates that the two variables are negatively related. Unfortunately, a correlation coefficient does not have a substantively meaningful unit. Instead correlation coefficients are explained based on general benchmarks indicating the strength of the relationship. A coefficient of –.1525 would typically be explained as suggesting a weak negative relationship. A value closer to –1 would suggest a stronger, negative relationship.

Although the number of activities is an important factor affecting the number of hours a young adult may work, surely there are other variables that may also be important. For example, you might argue that young adults who believe that marriage should ideally wait until they are older would work less hours because they are not ready to be or do not think they are fully adults (i.e., they try to limit their participation in the full-time "adult" work force). The NSYR data have a variable, `marrymin`, that comes from a question that asked respondents what they believed the ideal age to get married is. The variable is named `marrymin` because it holds either the respondents' exact ideal age of marriage or the minimum age if a range was provided.

Ideally, you would follow all the steps given above to examine this new variable. But for illustration purposes, jump directly to determining the correlation coefficient between `workhrs1` and `marrymin`. To do so, you do not need to enter a completely new `-corr-` command, rather you can simply add the new variable to end of the previous, bivariate command.

Type `corr workhr totacts marrymin` into the Command window (or use the **Page Up** button and add the new variable to the end of the last run command) and press **Enter** to display the following table:

```
(obs=2459)

             | workhrs1  totacts marrymin
-------------+---------------------------
    workhrs1 |   1.0000
     totacts |  -0.1508   1.0000
    marrymin |  -0.0197   0.0180   1.0000
```

Just as before, the matrix display shows each of the possible coefficients. You are primarily interested in the two listed within the `workhrs1` column, but the correlation between `totacts` and `marrymin` is also shown.

Before interpreting these values, you might notice two important differences between this table and the previously produced correlation table. First, the correlation coefficient between the `totacts` and `workhrs1` variables has decreased slightly. This change seems to suggest that including the `marrymin` variable somehow changes the calculation of the correlation coefficient between work hours and total activities. This discrepancy, however, is not due to the calculation being different (as the `-corr-` command only computes all the possible bivariate correlation coefficients), but rather stems from the second difference in the table—the number of observations. The number of observations used in this analysis is 2,459, whereas in the previous correlation table it was 2,515.

This change is the result of the -corr- command automatically employing what is referred to as "listwise deletion." As was shown in the -sum- output above, each variable suffers from a few missing cases. For example, a few people claimed to not know the number of hours they work in a typical week. When the -corr- command is executed, Stata calculates each of the correlation coefficients using only cases that have responses to *all* the variables listed in the -corr- command line (i.e., "deleting," for the calculation of the correlation coefficient—not from the actual data, all the cases with missing information in the list).

Listwise deletion is one way to handle missing data. Some researchers prefer to employ what is called "pairwise deletion" when calculating correlation coefficients. Instead of removing all the cases with missing information on any variable in the list from the calculation, pairwise deletion only removes cases with missing information on the two variables used to calculate each correlation coefficient.

Rather than using an option, Stata has a separate command for calculating correlations using pairwise deletion. -pwcorr- automatically uses all the valid cases for each possible correlation coefficient calculated. A useful option to employ with -pwcorr- is -obs-, which lists how many cases were used for each calculation.

Type `pwcorr workhr totacts marrymin, obs` into the Command window and press **Enter**.

```
             | workhrs1  totacts marrymin
-------------+---------------------------
    workhrs1 |   1.0000
             |     2527
             |
     totacts |  -0.1525   1.0000
             |     2515     2516
             |
    marrymin |  -0.0228   0.0180   1.0000
             |     2470     2460     2471
             |
```

The output is very similar to the table produced by the -corr- command, except now the observations are listed under each coefficient. You can see that the correlation between the workhrs1 and totacts variables is the same as it was in the bivariate -corr- command run above because Stata used the same cases (2,515) to calculate both coefficients.

The table also shows that the marrymin variable contains a few more missing cases than the other variables, mainly because several people said they did not know their ideal age of marriage. The correlation between the marrymin variable and workhrs1 is negative, showing that, as predicted, younger adults who believe marriage should wait until an older age tend to work fewer hours. The value of this coefficient (-.0228), however, suggests that this relationship is very weak and is considerably weaker than the relationship between total work hours and number of extracurricular activities.

The -pwcorr- command also contains an option, -sig-, to examine the significance of the correlations. The -corr- command does not have this option, which is another reason the -pwcorr- command may be preferred. Type pwcorr workhr totacts marrymin, obs sig into the Command window and press **Enter**.

```
             | workhrs1  totacts marrymin
-------------+---------------------------
    workhrs1 |   1.0000
             |
             |     2527
             |
     totacts |  -0.1525   1.0000
             |   0.0000
             |     2515     2516
             |
    marrymin |  -0.0228   0.0180   1.0000
             |   0.2580   0.3710
             |     2470     2460     2471
```

Notice that none of the correlations have changed. But the line underneath each of the correlations now presents their associated *p* value. The *p* value for the correlation between number of work hours and total activities is .000, providing evidence for a significant relationship between the two variables. Conversely, the *p* value for the correlation between the number of work hours and the ideal age of marriage is .2580, which is greater than a standard a level of .05, meaning you would fail to reject the null hypothesis and conclude that the correlation coefficient is not significantly different from 0.

Linear Regression

One of the major drawbacks of the correlation coefficient is its lack of meaningful units. The benchmarks used to determine the strength of a correlation coefficient are somewhat vague and do not provide an interpretation in terms of the measure under study (e.g., How much do work hours decrease?). A linear regression analysis is a common analytic technique to examine the relationship between two interval-ratio variables that allows for a more substantive interpretation.

To understand how a linear regression analysis is performed in theory, it is helpful to return to the image of the scatterplot. The scatterplot shows the general form of the relationship between two interval-ratio variables. A regression analysis attempts to summarize that relationship by drawing a line that best illustrates the direction of the points. This line is drawn, often referred to as the "best fitting line," in a way that minimizes the distance between each of the points and itself. You can imagine drawing numerous different lines that are angled slightly differently and then calculating how far each point is from each line. The line that produces the smallest total difference would be considered the best fitting line. The regression analysis then provides the equation, including the intercept and the slope, that is used to construct that best fitting line. In reality, a linear regression analysis uses mathematical formulas to determine the equation for the best fitting line, but this visual description may help make its analytic motivation easier to understand.

If you were going to ask a smart colleague to conduct a linear regression analysis, the most intuitive one-word directive you might use would be `-regress-`. The `-regress-` command, generally abbreviated as `-reg-`, produces an ordinary least squares linear regression. The basic structure for this command is fairly straightforward and very similar to the one used to produce a correlation coefficient. The primary difference is that a regression analysis is not a symmetric technique, meaning that the variable designated as the dependent variable versus those set as the independent variables makes a difference. Therefore, the ordering of the variables in the command line matters. If you have been following the previous suggestions about variable ordering in analytic commands, this pattern should be easy to keep straight. The first variable entered after the command must be the variable that is being treated as the dependent variable. The independent variable(s) are typed next. Therefore, the basic form of the `-regress-` command is

```
reg DV IV1 IV2 IV3 ... IVn
```

Notice that the form of the command is very similar to the way that a linear regression equation is typically expressed: $y = bx_1 + bx_2 + bx_3 + \ldots + bx_n + a$. The main difference is that the equal sign and constant (a) are *not* needed and the *ys* and *bxs* are replaced with variable names. For readers who are unfamiliar with this notation, the subscript numerals indicate that you can enter as many, or as few, independent variables (*xs*) as desired (i.e., n). To start, however, try a basic bivariate linear regression (i.e., only include one independent variable).

In the example above, number of hours in paid employment is the dependent variable and total number of extracurricular activities is the primary independent variable. To conduct this bivariate regression, type `reg workhr totacts` into the Command line and press **Enter**. The following results are displayed:

```
      Source |       SS       df       MS              Number of obs =    2515
-------------+------------------------------           F(  1,  2513) =   59.84
       Model |  21110.4422     1  21110.4422           Prob > F      =  0.0000
    Residual |  886551.667  2513  352.786179           R-squared     =  0.0233
-------------+------------------------------           Adj R-squared =  0.0229
       Total |  907662.109  2514  361.043003           Root MSE      =  18.783

------------------------------------------------------------------------------
    workhrs1 |      Coef.   Std. Err.      t    P>|t|     [95% Conf. Interval]
-------------+----------------------------------------------------------------
     totacts |  -1.297177    .1676897    -7.74   0.000    -1.626002   -.9683532
       _cons |   22.47087    .4553436    49.35   0.000     21.57798    23.36376
------------------------------------------------------------------------------
```

Before moving into the substantive interpretation, it can be helpful to identify the components of the output produced by the `-reg-` command. The upper portion of the results displays the results for the overall regression equation. The top left portion lists the figures associated with the sum of squares portion of a linear regression analysis. `Model` refers to what is sometimes referred to as the regression sum of squares and `Residual` represents what is often called the error sum of squares. These statistics are important to the regression equation but are rarely reported.

The upper right portion provides summary statistics for the entire regression equation. First, the number of observations included in the analysis is listed, followed by the calculated *F* statistic of model significance and its associated *p* value. The final three numbers shown are all measures of how well the equation fits the data (i.e., how effective the independent variables are at predicting the dependent variable), including the R^2, adjusted R^2, and the root mean square error.

Below these overall regression statistics, the information for each independent variable and the constant (sometimes referred to as the intercept) are listed. Moving from left to right, the output displays the coefficients, its standard error, the calculated *t* value, *p* value for that *t* value, and the 95% confidence interval for the coefficient.

The figure of primary importance is the one listed underneath `Coef` in the `totacts` row. This statistic is typically called the beta coefficient, and in a bivariate linear regression analysis, this represents the slope of the best fitting line. Similar to the correlation coefficient, the sign of the beta coefficient indicates the direction of the relationship between the `totacts` and `workhrs1` variables. As above, the negative beta coefficient suggests that as the number of activities increases, the number of hours spent in paid employment should decrease. Unlike a correlation coefficient, a beta coefficient can theoretically range from positive to negative infinity, but larger, positive or negative, values still indicate a stronger relationship.

The beta coefficient itself can be interpreted as how much the dependent variable changes based on a one-unit increase of the independent variable. In this example, a coefficient of −1.30 indicates that for each additional activity a young adult is involved in, he or she would be predicted to work just over one hour less per week. This type of interpretation gives a very clear picture of how strongly the two variables are related and is one of the primary strengths of linear regression analyses.

The following columns in the `totacts` row present statistics for determining the significance of the beta coefficient. The significance test of a beta coefficient is based against the assumption that the coefficient is 0 or that there is no relationship between the two variables. The *p* value of .000 in the example, listed under the column heading `P>|t|`, is lower than the standard a of .05, which would mean you would reject the null hypothesis and conclude that the beta coefficient is significantly different from 0.

The next row, `_cons`, provides information of the regression constant or intercept. The coefficient for the constant is the starting point of the best fitting line. This starting point is always based on where the line would cross the Y-axis, meaning that the independent (X) variable(s) would be equal to 0. Therefore, the constant coefficient can be interpreted as the predicted value of the dependent variable when the independent variable is 0. The example indicates that when young adults do not participate in any extracurricular activities, they should work almost 22.5 hours per week.

The final pertinent statistics are reported in the upper right portion of the display. These figures provide information concerning how well all the independent variables entered into the analyses do at predicting the dependent variable. Perhaps the most useful figure for this purpose is the `R-squared` value, sometimes referred to as the coefficient of determination. The `R-squared` value can range from 0 to 1, with higher values showing that the independent variables do a better job of predicting the outcome. The specific value can be interpreted as the percentage of the variation in the dependent variable that is accounted for by the independent variable(s). In this example, the `R-squared` value is .023, meaning the 2.3% of the variation in number of work hours is accounted for by the total number of extracurricular activities. When only using one predictor variable, the `R-squared` value is literally the correlation coefficient (i.e., *r*) squared. Using the correlation table shown above, you can see that $(-.1525)^2$ indeed equals .023. This particular `R-squared` is relatively weak, although for including only one variable in the analysis, it is not terrible.

MULTIPLE LINEAR REGRESSION

Chapter 5 introduced the elaboration method to account for possible confounding third factors when determining the relationship between two ordinal variables. Remember the idea was that you examine the relationship between the dependent and one independent variable at specific levels of the third variable. This general strategy can be carried over to the analysis of interval-ratio variables through the use of multiple linear regression.

Multiple linear regression is very similar to a bivariate linear regression, except that more independent variables are included in the prediction. In the example above, you considered that young adults perceived ideal age of marriage might influence how many hours they work. Potentially, the relationship between activities and work hours stems from the spurious relationship of both variables with the ideal age of marriage. Young adults who believe that marriage should occur at an older age might believe that adolescence should be extended as long as possible, which may cause them to work less and participate more heavily in extracurricular activities. This spurious relationship could cause the observed negative relationship between extracurricular participation and number of work hours. To test this possible confounding relationship, the perceived ideal age of marriage needs to be "controlled" for in the linear regression analyses.

Including multiple independent variables in a regression equation is simply a matter of entering them at the end of the `-reg-` command line. The `-reg-` command allows for essentially as many independent variables as you might have in a data set, and the order of the independent variables does not matter for the estimation of each variable's regression coefficient.

Type `reg workhr totacts marrymin` in the Command window and press **Enter**. The results now appear as follows:

```
      Source |       SS       df       MS              Number of obs =    2459
-------------+------------------------------           F(  2,  2456) =   28.94
       Model |   20269.981     2  10134.9905           Prob > F      =  0.0000
    Residual |  860253.311  2456  350.266006           R-squared     =  0.0230
-------------+------------------------------           Adj R-squared =  0.0222
       Total |  880523.292  2458   358.22754           Root MSE      =  18.715

------------------------------------------------------------------------------
    workhrs1 |      Coef.   Std. Err.      t    P>|t|     [95% Conf. Interval]
-------------+----------------------------------------------------------------
     totacts |  -1.270075   .1683788    -7.54   0.000    -1.600254   -.9398962
    marrymin |  -.0995099   .1169384    -0.85   0.395     -.328818    .1297982
       _cons |   24.95128   3.019792     8.26   0.000     19.02967    30.87288
------------------------------------------------------------------------------
```

The basic layout of the results is identical to the bivariate regression produced above.

To analyze the potential spuriousness of the ideal age of marriage variable, refer to the coefficients and *p* value of both the `totacts` and `marrymin` variables. If the perceived ideal age of marriage is a spurious factor, the coefficient of the `totacts` variable should be equal to 0 or greatly diminished when the analysis includes both variables. The coefficient for total activities (−1.27), however, remains virtually unchanged and the *p* value of .000 is still less than the standard a level of .05, meaning the coefficient is significantly different from 0. Therefore, as young adults' extracurricular participation increases, their work hours should decrease, even when controlling for the perceived ideal age of marriage.

Furthermore, the coefficient for the `marrymin` variable is very small (-.1). For each year older a young adult believes is the ideal age of marriage, his or her work hours would be predicted to only decrease by about 6 minutes (-.1*60 minutes). The *p* value `(P>|t|)` for the `marrymin` variable (.395) is greater than the standard a level of .05, meaning you fail to reject the null hypothesis and must conclude that the true regression coefficient for ideal age of marriage is not significantly different from zero. Finally, the `R-squared` value has remained virtually unchanged, which suggests that including ideal age of marriage in the analyses does not improve the prediction of hours of employment.

A Closer Look: Predictions After Regression Analyses

A comprehensive regression analyses often requires what are referred to as "diagnostics," which are postregression tests that examine the adequacy and validity of the analyses. A full explanation of diagnostics is beyond the scope of this book, but virtually all these tests require values to be calculated based on the regression equation that was estimated in the regression analyses. Generally, these values are called predictions. The `-predict-` command can be used after the `-reg-` command has been invoked to calculate various predictions.

The basic structure of the `-predict-` command is

```
predict newvarname, predvalopt
```

(Continued)

(Continued)

`newvarname` is where you tell Stata the name of the variable that will hold whatever prediction is specified by the `predvalopt`.

Two of the most commonly used predictions are predicted values and residuals (often referred to as the standard error of the prediction). Predicted values use the estimated beta coefficients to calculate the predicted value of each case on the dependent variable based on its score on the independent variable(s). Residuals are the difference between this predicted value and the actual value of each case on the dependent variable. The option to specify that predicted values should be calculated is `-xb-`, while to generate residuals the option is `-stdp-` (standing for standard error prediction).

After the previous regression command (`-reg workhr totacts marrymin-`) has been invoked, the following two commands would produce the predicted values and residuals.

```
predict predvals, xb
predict resids, stdp
```

The variable `predvals` would contain the predicted value for each case based on the previously run regression analysis. Similarly, the variable `resids` would contain the residual value of each case. These variables could then be used to conduct various diagnostic tests (such as a scatterplot of the two).

Predicted values and residuals are only two of the many values that can be produced by using the `-predict-` command after a regression analysis. Chapter 8 explains ways to learn the specific option names for each possibility.

DICHOTOMOUS (DUMMY) VARIABLES AND LINEAR REGRESSION

Thus far this chapter has focused exclusively on prototypical interval-ratio variables. There is one type of variable that can be used in linear regression analysis that is not exactly an interval-ratio variable. Variables with only two categories, typically referred to as dichotomous or "dummy" variables, are a special type of variable. Although they have limited categories, they can be treated, in practice, as interval-ratio variables. Examples of this type of variable are gender, being married or not, and having a high school diploma or not. The specifics behind how these types of variables are treated statistically in a linear regression analysis are beyond the scope of this book. The way in which they are included in such analyses in Stata, however, is very straightforward.

An example of a dichotomous factor that might matter for the analysis of work hours is whether the respondent is currently dating or not. Young adults who are in romantic relationships might have to work longer hours to support the types of behaviors typically associated with dating (e.g., going to dinner, car payments, and buying gifts). This relationship should not vary by how long two people have been dating or how many people someone has dated, making the dichotomous, dating or not, variable the pertinent measure. The NSYR data contain a variable, `dating`, based on a question about whether the respondent was currently in "a dating or romantic relationship." The frequency distribution, including cases coded as missing, of this variable (`-tab dating, mis-`) is listed below.

```
(dating_w3) [IF |
         IS NOT |
      CURRENTLY |
MARRIED AND HAS |
      BEEN IN A |
       ROMANTIC |
  RELATIONSHIP] |      Freq.     Percent        Cum.
----------------+-----------------------------------
             No |        955       37.72       37.72
            Yes |      1,254       49.53       87.24
             .d |          1        0.04       87.28
             .r |          3        0.12       87.40
             .s |        319       12.60      100.00
----------------+-----------------------------------
          Total |      2,532      100.00
```

The variable is indeed dichotomous, with No and Yes being the only non-missing categories. The table shows that about 50% of all respondents report being in a dating relationship. It also shows that there is a large portion (12.6%) of cases that were skipped out of this question. These cases were skipped because they had previously reported that they had *never* been in a dating or romantic relationship. Because you are assessing whether being in a current dating relationship or not is related to work hours, it would be defensible to recode the skipped cases (i.e., have never dated) on this variable as not currently being in a relationship (i.e., as 0).

Type `recode dating (.s=0), gen(currdate)` into the Command window and press **Enter**. Then create a frequency distribution of the newly created `currdate` variable by typing `tab currdate` into the Command window and pressing **Enter**.

```
  RECODE of |
     dating |
 ((dating_w3 |
   ) [IF IS |
        NOT |
  CURRENTLY |
MARRIED AND |
HAS BEEN IN |
      A ROM |      Freq.     Percent        Cum.
------------+-----------------------------------
          0 |      1,274       50.40       50.40
          1 |      1,254       49.60      100.00
------------+-----------------------------------
      Total |      2,528      100.00
```

The 319 skipped (`.s`) cases have been added to the 955 No cases, leading to a total of 50.4% of respondents reporting that they are not currently in a dating relationship.

The `-recode-` command above made the assumption that No is coded as 0 on the dating variable. This assumption is justified because all "Yes/No" questions in the NSYR data are coded as *No* = 0 and *Yes* = 1. This default coding strategy may not be the case in all data sets, which is why the `-tab-` command with the `-nol-` option should always be used to double-check the coding of categories. Dichotomous variables do not have to be coded as 0 and 1 to operate properly in the `-reg-` command. They could be coded 1 and 2 or even 100 and 101. The only rule that dichotomous variables must follow is that the categories must be coded as consecutive integers.

In addition to including this new dichotomous variable into the regression analysis, it may be helpful to compute the standardized beta coefficients. The interpretations of unstandardized beta coefficients are made in terms of the units of the particular variable (e.g., one additional activity or 1 year older). This lack of standardized units makes it impossible to compare the magnitude of beta coefficients across independent measures. Standardized beta coefficients convert the actual units of a variable into a standard unit (using standard deviations). These coefficients can then be used to compare the relative strength of the independent variables' relationship with the dependent variable. The option used to display the standardized beta coefficients is `-beta-`.

To include this new dichotomous variable in the regression analysis and display the standardized beta coefficients, type `reg workhr totacts marrymin currdate, beta` into the Command window and press **Enter**.

```
      Source |       SS       df       MS              Number of obs =    2455
-------------+------------------------------           F(  3,  2451) =   22.72
       Model |  23786.383      3  7928.79435           Prob > F      =  0.0000
    Residual | 855232.821   2451    348.9322           R-squared     =  0.0271
-------------+------------------------------           Adj R-squared =  0.0259
       Total | 879019.204   2454  358.198535           Root MSE      =   18.68

------------------------------------------------------------------------------
    workhrs1 |      Coef.   Std. Err.      t    P>|t|                     Beta
-------------+----------------------------------------------------------------
     totacts |  -1.246716   .1682522    -7.41   0.000                -.1477701
    marrymin |  -.0689596   .1171517    -0.59   0.556                -.0117715
    currdate |   2.429522    .757317     3.21   0.001                 .0641973
       _cons |   22.91394   3.081775     7.44   0.000                        .
------------------------------------------------------------------------------
```

The main difference in the output is that the standardized beta coefficients, labeled Beta, have taken the place of the confidence intervals of the coefficients.

The interpretation of the currdate variable is very similar to the ones made previously for the totacts and marrymin variables. Instead of talking about when dating "increases," however, the interpretation is made in terms of young adults dating versus not dating. That is, the coefficient of 2.43 indicates that a young adult who is dating is predicted to work almost 2.5 more hours than a young adult who is not currently in a dating or romantic relationship, when controlling for total extracurricular activities and perceived ideal age of marriage. Including this variable in the analysis has improved the overall prediction of work hours. The R-squared value shows that the three variables collectively account for 2.7% of the variation in work hours.

Examining the Beta column reveals that the total number of activities has the strongest relationship of the three with work hours. The absolute value of the standardized beta coefficient for totacts is greater than both marrymin and currdate. Although somewhat tentative, you can conclude that of these three variables, the total number of activities in which a young adult is involved has the largest influence on the number of hours he or she works in a typical week.

Summary of Commands Used in This Chapter

```
gen totacts=relact+notrelac
sum workhr totacts, det

*Scatterplots
scatter workhr totacts
scatter workhr totacts if totacts<20
```

```
*Correlation
corr workhr totacts
corr workhr totacts marrymin
pwcorr workhr totacts marrymin, obs
pwcorr workhr totacts marrymin, obs sig

*Regression
reg workhr totacts
reg workhr totacts marrymin
tab dating, mis
recode dating (.s=0), gen(currdate)
tab currdate
reg workhr totacts marrymin currdate, beta

*A Closer Look: Predictions After Regression
Analyses
reg workhrs totacts marrymin
predict predvals, xb
predict resids, stdp
```

Exercises

Use the original `Chapter 7 Data.dta` for the following problems. [Optional: Complete the exercises by using a do file and save the results using a log file. See Chapter 3 for an explanation of how to use these files.]

1. Create a scatterplot to look at the relationship between the number of religious retreats young adults attend (`relretrt`) and the number of children they want (`kidwntmn`). [Note: For this example treat the number of desired children as the dependent variable.]
2. Alter the scatterplot between the `kidwntmn` variable and `relretrt` variable so that it only includes respondents who report attending less than 20 religious retreats.
3. Display the listwise Pearson's correlation coefficient for the `kidwntmn` variable and the `relretrt` variable.
4. Produce the pairwise Pearson's correlation coefficient for the `kidwntmn` variable, the `relretrt` variable, and the ideal age of marriage (`marrymin`) variable. Also display the observations used for each calculation.
5. Conduct a bivariate regression analysis to investigate whether attending religious retreats leads to a difference in desired number of children.
6. Use a regression analysis to examine whether the ideal age of marriage alters the bivariate relationship between religious retreat attendance and number of children desired.
7. Include the dichotomous `gender` variable into the previous regression analysis and display the standardized regression coefficients to assess the relative influence of each variable on the number of children young adults want.

8

Enhancing Your Command Repertoire

The most difficult aspect of writing a book such as this is determining which topics to cover. Given the vast diversity of research interests and quantitative analysis strategies, it is impossible to write a manageable introductory text without excluding some commands that many readers believe to be important. Hopefully, for new researchers and Stata users, this book has minimized that critique.

As you become more proficient in both analytic techniques and Stata commands, it is inevitable that you will require a command that accomplishes some task that has not been covered here. Rather than writing a Stata tome, it is more advantageous to learn a few basic strategies for learning how to help yourself. This chapter first addresses the use of Stata help files, which, along with the manuals, can be extremely beneficial in learning the vast capabilities of the program. Second, it presents a few supplementary commands. In addition to illustrating the specific utility of these commands, this section is presented to prove that once you have the basic knowledge of how to use the Stata commands presented in the previous chapters, learning new commands is not nearly as difficult as it may have at first seemed. And enhancing your command "tool kit" will continue to increase Stata's effectiveness for your research (and possibly even your enjoyment).

Stata Help Files

Perhaps one of the most common complaints of any computer software program is that its built-in help files are not very helpful. Assuredly, more than a few new users have proffered this criticism against Stata. This dissatisfaction, however, mainly stems from the unique vernacular and structure of Stata's help files. Don't worry, you do not have to learn a new language to comprehend Stata's help files. Understanding a few simple codes unlocks the utility of Stata's self-help devices. The following section is geared to help you learn how to use and interpret Stata's help files so that rather than being a source of annoyance, they can be a tool that assists you in completing your analyses and furthering your Stata capabilities.

WAYS TO SEARCH AND ACCESS

There are four primary ways that you can find a help file. Each takes a different path to reaching the same goal. The first procedure is to click on the **Help** menu button at the top of the main Stata window and then click on **PDF Documentation.** Doing so will open a PDF file that contains the full Stata manual. It is organized first by topic and then alphabetically by command name. You can use the navigation on the left side of the window to find a particular topic, and each of the command names are "clickable," taking you directly to that particular help file.

The second option for accessing help files is very similar but is organized in more of a web-based interface. Again, click on the **Help** menu button and then click on **Contents.** The following window appears:

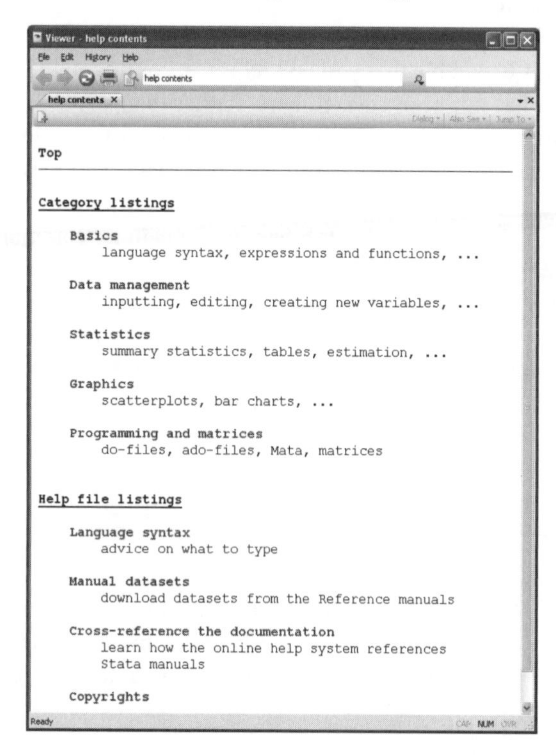

Each of the headings can be clicked, which will open another window with more detailed headings. You can keep clicking through the headings until you find the help file you are looking for.

Both of the first two methods are most effective if you are looking for broad categories of commands. For instance, if you wanted to know all the possible types of regression analyses that Stata can conduct, you could use the above method and then click Statistics→ Estimation→Regression Models and the window below is shown:

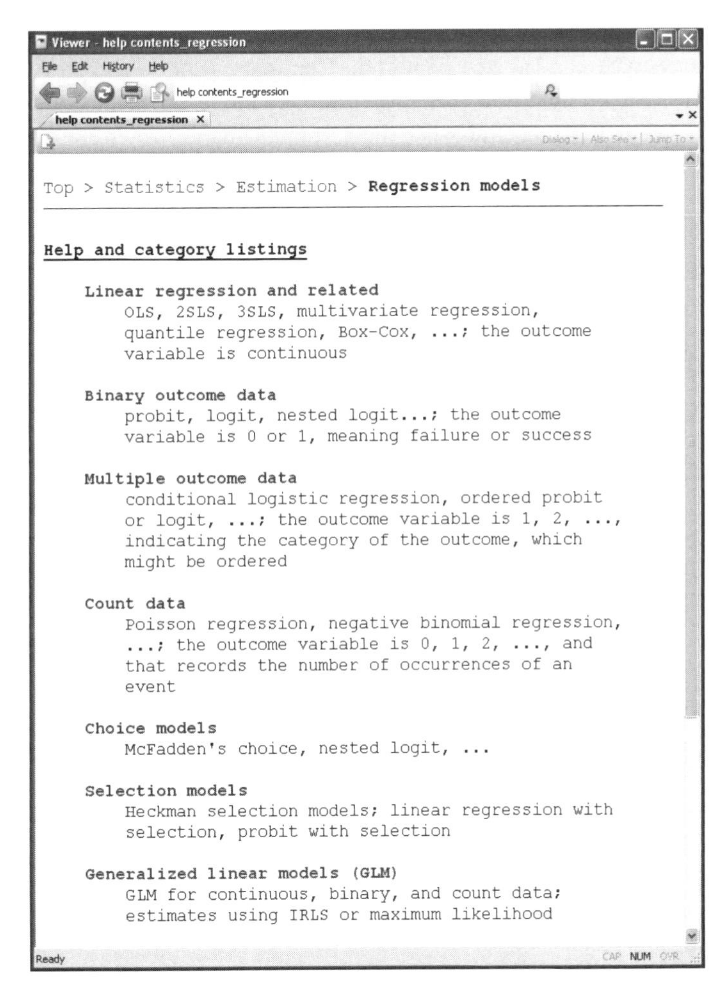

You can then click on the appropriate type of regression you are interested in, and the necessary commands and their help files would be displayed.

Many times, you may be searching for help with a slightly more specific goal in mind. For example, imagine that this manual had not covered how to produce a Pearson's correlation coefficient. You could use these first two methods to peruse through the various headings to find correlations, or you can take a slightly more direct route.

The third method for finding help files starts by clicking on the **Help** menu and then clicking **Search**, which brings up a search engine window. You can type in a topic and Stata can search its own help file and/or resources on the Internet.

The fourth method is similar but even slightly quicker because it uses the Command window interface. The command to display a help file is `-help-` followed by the topic you are searching for. If you know the command name but want to know more about it (e.g., the full list of options that are available), you can enter the command name directly after `-help-` and the exact help file is shown (e.g., type `-help corr-` into the Command window and press **Enter**).

When you do not know the exact command name, however, you can simply enter your best guess at what the command name would be or even a general topic. If you did not know that `-corr-` was the command name to produce a Pearson's correlation coefficient, and were thinking intuitively, you might first try "pearson."

Type `help pearson` into the Command window and press **Enter**. The following window is displayed:[1]

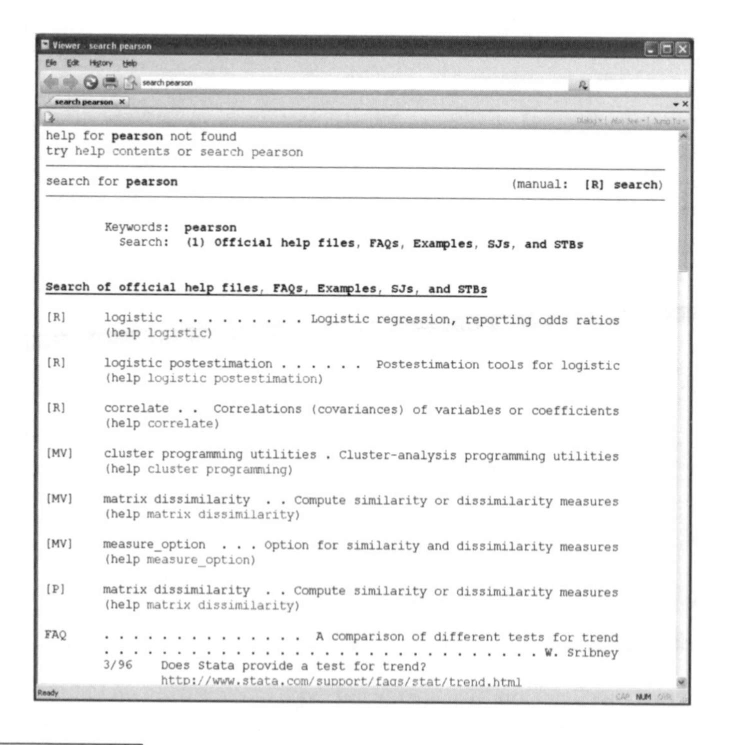

[1]In versions prior to Stata 12, an initial dialog box will appear asking whether you would like to perform a keyword search on the entered topic (e.g., "pearson"). Click on Yes or type Y, and a window similar to the one shown will be displayed.

This "search" box only indicates that there is not a command that exactly matches "pearson." It does not mean that a command does not exist that will produce a Pearson's correlation coefficient.

The window displays the results of Stata's keyword search on "pearson." Several different types of resources are searched and displayed, including help files, FAQs, examples, and even *Stata Journal* articles. Any listing that has the `[R]` symbol to the left is a help file directly from the Stata manual. When you find the one you are looking for, in this case it is the third listing, you can click on the command name (listed in parentheses after the word "help"), and the help file is displayed.

STRUCTURE AND LANGUAGE

To examine an actual help file, you can simply click on `correlate` from the search results window (if you were following the example from above). Or type `help corr` in the Command window and press **Enter**.

```
Viewer - help correlate
File  Edit  History  Help
help correlate
help correlate ×
Dialog ▾  Also See ▾  Jump To ▾

Title

    [R] correlate — Correlations (covariances) of variables or
                         coefficients

Syntax

    Display correlation matrix or covariance matrix

        correlate [varlist] [if] [in] [weight] [, correlate_options]

    Display all pairwise correlation coefficients

        pwcorr [varlist] [if] [in] [weight] [, pwcorr_options]

    correlate_options     Description
    ------------------------------------------------------------------
    Options
      means               display means, standard deviations, minimums, and
                            maximums with matrix
      noformat            ignore display format associated with variables
      covariance          display covariances
      wrap                allow wide matrices to wrap
    ------------------------------------------------------------------

    pwcorr_options        Description
    ------------------------------------------------------------------
    Main
      obs                 print number of observations for each entry
      sig                 print significance level for each entry
      listwise            use listwise deletion to handle missing values
      casewise            synonym for listwise
      print(#)            significance level for displaying coefficients
Ready                                                       CAP NUM OVR
```

This window shows a standard Stata help file. Again, it may seem daunting at first, but after you become accustomed to the layout and lingo, it will be extremely advantageous.

The basic organization is the same for all help files. Under `Title` is the command name and a brief description of the operation that command performs. Then under `Syntax` the structure of the command(s) is presented. This "syntax" line provides information on what must and can be typed into the Command window to execute the command. Next the command's options and a description of their functionality is shown. Additionally (not shown in this screenshot), some basic notes on the restrictions of the command are described, such as whether it can be used with the `-by-` prefix command. In this example, you can see that you can use the `-by-` prefix with both commands, as well as two types of weights.

Throughout the help file different fonts represent different aspects of the command. Any word that is presented in bold font is something that can be typed into the Command window, usually the command name and its options. Many of these bold words have a piece of their name underlined. The letters that are underlined represent the minimum portion of the command or option that can be typed in the Command window. For example, the full command `-correlate-` can be abbreviated as `-corr-` and the option `-mean-` could be typed as just `-m-`.

Words that are shown in a different color, typically blue, can be clicked to open their own, dedicated help file. Anything that is listed in italics is something that the user must "fill in" when typing the command. You should not type the word that is in italics. Rather it provides an indication of what should be typed in that location of the command. Additionally, portions of the command that are listed in brackets ([ ]) are not required. For example, with `-corr-` you do not have to type anything after the command name, shown by every aspect being in brackets. If you were to type `corr` into the Command window and press Enter, Stata would calculate and display the correlation coefficient for every single possible combination of the variables in the data set. Typically, something, usually variable names, needs to be typed in addition to the command name, but remember if it is in brackets, it is completely optional.

Perhaps the most crucial and also the least straightforward aspect of help files is the portion of the syntax listed directly after the command name. In the `-corr-` help file, this portion reads *`varlist`*. The word *`varlist`* indicates that you can type a list of variable names after the `-corr-` command. The most common codes you will see in this portion of a help file are some variant of *`var`* (e.g., *`newvar`*, *`varname`*, *`depvar`*, referring to a new variable name, existing variable name, and a dependent variable, respectively) and =*`exp`*. The latter refers to an "expression," which is some type of formula involving values and/or variable names (e.g., the -gen- command requires an expression after the new variable name has been typed). In addition to the syntax line, some of these codes appear in the options. For example, the option -level(#)- contains the code #, which means you need to enter a number.

At first these codes may seem opaque, but the more you use help files, the clearer they become. It can be helpful to examine the help files of commands that you already know as a method for learning how to understand the help files of commands with which you are less familiar. Additionally, most of these codes are clickable, taking you to a help file further explaining their meaning. An abbreviated list and explanation of the most common codes is listed in the "A Closer Look" box below.

A Closer Look: Help File Code Words

The following table provides an abbreviated list of the language (or codes) that commonly appear in help files as well as a description of how they should be interpreted. It also provides an example of a command, used previously in this book, with the portion of the specified language in bold typeface.

Language/ Code	Interpretation	Example
varname	The name of a variable in the data set	tab **employst;**
varlist	A list of variables in the data set	corr **workhrs totacts marrymin**
newvar	The name of a new variable	gen **agep16**=agecats-16
indepvars	The name of variables in the data set that are specified as independent variables	reg workhrs **totacts marrymin**
depvar	The name of a variable in the data set that is specified as the dependent variable	reg **workhrs** totacts marrymin

(Continued)

(Continued)

Language/ Code	Interpretation	Example
=exp	An expression. Expressions usually involve some type of formula involving a combination of mathematical operators and/or variable names	`gen agep16=`**`agecats-16`**
#	A number	`ci freqvol, level`**`(99)`**

If you scroll down the help file window, a more thorough description of the operation the command performs is provided, as well as much more information on what the options do. Finally, and perhaps most helpfully, you notice the following portion of the help file toward the bottom of the window:

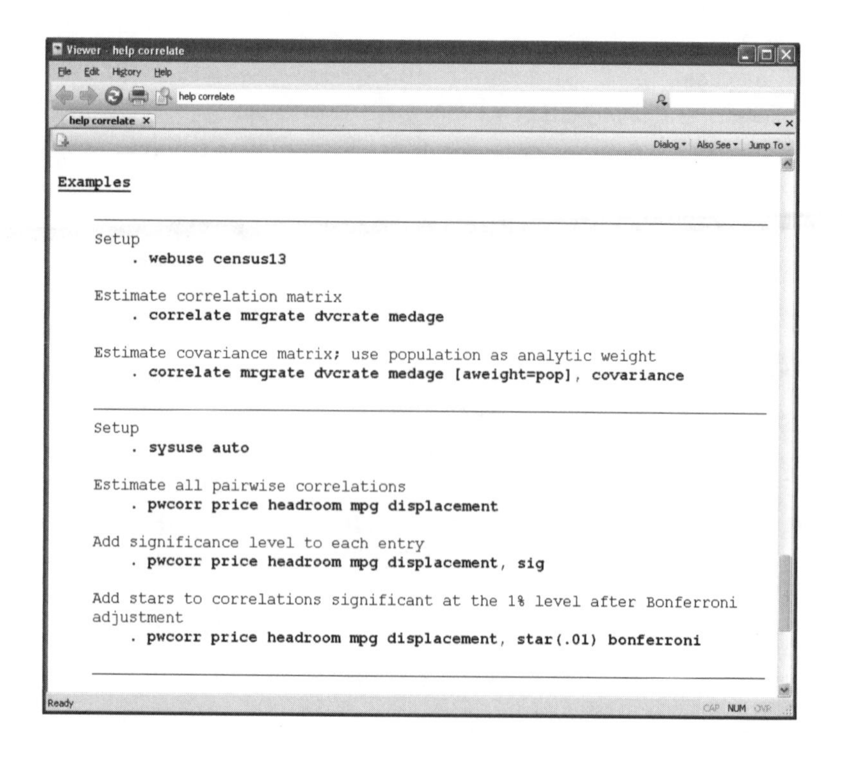

These examples show how exactly to use the command, including its various options. All the lines that are preceded by a "." are commands that can be typed into the Command window. Furthermore, these examples use easily accessible data so that you can practice the different commands to see what type of results they display. The first line of the example is a command to open the data set used for the given examples. (Note: You must be working on a computer that is connected to the Internet to open data that are opened with the `-webuse-` command. Data opened with the `-syuse-` command can be opened without an Internet connection.)

Advanced Convenience Commands

The following section explains a few additional commands that can be thought of as "convenience commands." They are called convenience commands because all their operations can be accomplished using commands that have already been discussed. These commands are simply shortcuts.

They are "advanced" because their structure is a bit different from previously discussed commands. First-time users introduced to these types of commands can start to have some frustration as it may begin to seem as if there are endless commands that are all slightly different in form. If you start to sense this irritation building, remember you can use the commands you have already learned to reach the exact same end-product of these commands. It may even be advisable for first-time users to become confident and comfortable with the commands that have already been covered before diving into this section. Of course, if this manual has accomplished its goal of helping minimize that sense of being overwhelmed, then by all means dive away.

All the examples that follow use the `Chapter 8 Data.dta`, available at **www.sagepub.com/longest**. This data set includes several variables from the National Study of Youth and Religion (NSYR) data that have been used in previous chapters and a few new measures. It still contains the full Wave 3 sample of 2,532 young adults. All the missing cases have been replaced with appropriate missing codes in this data set, with `.d` referring to a response of "Don't Know," `.r` to a response of "Refused," and `.s` to a case that was legitimately skipped out of a question based of the survey design (i.e., skip pattern).

tab, gen(newvar)

The first convenience command is not necessarily a new command, but rather a new option. In several previous chapters, the `-tab-` command has been used to display frequency distributions and cross-tabulations. These two operations are its primary purpose, but the command allows for a `-gen(newvarstub)-` option that can be quite useful.

Chapter 7 discussed using dichotomous, "dummy" variables in regression analyses. In addition to a single dichotomous variable, a set of dummy variables, based on the categories of a nominal variable, can be used as predictors in regression analyses.

For example, in Chapter 6, the nominal variable representing religious denomination affiliation, i_religi (renamed denom), was examined in relation to frequency of volunteering. This variable contains six different categories: Evangelical Protestant, Mainline Protestant, Black Protestant, Catholic, Not Religious, and Other Religion. To use this variable as a dummy set in a regression analysis, each of these categories would need to have its own single, dichotomous indicator variable. One of the categories, such as Not Religious, would need to be selected as a "reference category," and only the remaining five dichotomous variables should be entered as independent variables into the regression analyses.

It would be possible to use a combination of -gen- and -replace- commands to create six individual, dichotomous variables. This process would be straightforward but would require several command lines. A shortcut is to employ the -gen(newvarstub)- option after the -tab- command. This option automatically produces a dichotomous variable, coded 0 and 1, for each of the categories of the variable entered after -tab-. The language newvarstub means that you have to provide a new "stub" (or root word) that is used to name the new variables. Each of the new variables share this common stub as the first part of their variable name, which is then followed by a number indicating the order in which that category appeared in the original, nominal variable. In the current example, this means that there will be six new variables created, all named with the provided stub followed by a number between 1 and 6, based on the order of the denomination categories in denom.

Type tab denom, gen(dumden) into the Command window and press **Enter**. The following table appears:

```
 (tradrel_w3) Identical |
  to relatt_w3 but uses |
  identification info on |
          non-attenders |      Freq.     Percent        Cum.
------------------------+-----------------------------------
 Evangelical Protestant |        714       28.20       28.20
    Mainline Protestant |        259       10.23       38.43
       Black Protestant |        189        7.46       45.89
               Catholic |        443       17.50       63.39
          Not religious |        622       24.57       87.95
         Other religion |        305       12.05      100.00
------------------------+-----------------------------------
                  Total |      2,532      100.00
```

In addition to the standard frequency table, you should see six new variables in the Variables window: `dumden1`, `dumden2`,... `dumden6`. Try producing a frequency distribution for `dumden2` (`-tab dumden2-`). It should appear as

```
denom==Main |
       line |
 Protestant |      Freq.     Percent        Cum.
------------+-----------------------------------
          0 |      2,273       89.77       89.77
          1 |        259       10.23      100.00
------------+-----------------------------------
      Total |      2,532      100.00
```

The results show that this variable is a dichotomous variable indicating being Mainline Protestant. The correct 259 cases are coded as 1, and the remaining cases are set to equal 0. It is named as `dumden2` because Mainline Protestant is the second category in the `denom` variable. Notice that the variable label, shown in the upper left corner of the table, helps clarify which category the particular variable represents, as well as the original variable from which it was derived.

A Closer Look: Using Wildcards

When you used the `-tab, gen()-` command, six separate variables with similar root names were created. You might be interested in seeing the frequency distribution for each variable. You could of course type `-tab dumden1 dumden2 dumden3 dumden4 dumden5 dumden6-` into the Command window. Clearly, this method is time-consuming. You can click on the arrow that appears next to each variable when you place your cursor over it in the Variable window to prevent having to type each specific variable name (or just the variable name itself if you are using Stata 11 or earlier). But there is an even quicker technique.

Stata allows for "wildcards" to be used when referencing variable names in commands. Wildcards are symbols that are used as an extreme form of shorthand. Perhaps the most frequently used wildcard is the asterisk. Placing an "*" after the root of a variable name tells Stata to perform the command on every variable that has that root, regardless of what comes after that root.

(Continued)

(Continued)

For example, you could type `tab1 dumden*` into the Command window and press **Enter**. Stata evaluates the wildcard and creates a frequency distribution for every variable in the data set that starts with the characters "`dumden`." In the current example, you could have even typed `dumd*`, as there are no other existing variables with this root.

The * wildcard can be used at the beginning of variable stubs or even in the middle. For example, if a data set contained variables called `denonevar dentwovar denthreevar`. You could type: `tab1 den*var`, and the frequency distribution of all the three variables would be displayed.

There is only a minimal danger when using wildcard with statistical analysis commands, as the worst scenario would be some variables might accidently be included in the analysis. If you are altering the data, however, caution is warranted. For example, if you wanted to `-drop-` the `dumden` variables, it would be important to make sure that there were no other variables with this root to prevent deleting variables accidentally.

One method for preventing this unintended consequence is to use the "–" wildcard. The – symbol is used in-between two variable names to tell Stata to perform the operation on all the variables in-between and including the two variables. "In between" in this context refers to the order of the variables in the data set, which can be seen in the Variables window. For example, the frequency distribution command shown above could also be entered as `tab1 dumden1 - dumden6`.

egen

As you might surmise, the `-egen-` command is used to generate new variables. It is a command that provides "extensions" to the `-gen-` command. And in many ways, its structure is exactly like the `-gen-` command. The primary difference is that instead of the user entering some type of formula in the expression portion of the command that follows the equal sign, the `-egen-` command has built-in functions that perform some of the most commonly needed calculations. The basic structure of the command is

```
egen newvar = fcn(arguments) [if] [, options]
```

The `fcn` portion of the command is where the commands are placed to designate the particular function to be performed on whatever is listed within the arguments parentheses. An example of a function is `-sd-`, which stands for standard deviation. If, for example, you wanted to create a new variable that contained the standard deviation of a particular variable, you could use the -sd- function and place the old variable within the parentheses. What is accepted or needs to be entered in the arguments section is specific to each function. The full set of functions can be seen by typing help egen in the Command window and pressing Enter.

Explaining the full set of possible functions could fill a chapter in its own right. But they all follow a similar structure, so two examples provide a sufficient introduction to the command.

Previously, you looked at the variable representing the body mass index (BMI) of respondents, called bmi. This variable contains the exact BMI calculation for each respondent. There might be situations where you would prefer to treat this interval-ratio variable as an ordinal variable. Currently, however, it has too many categories to do so. You could use the -recode- command to create a more condensed variable, but one of the functions in -egen- is specifically designed for this purpose.

The -cut- function cuts a variable either at specific values or into a set number of groups. Invoking the -at(#,#,...#)- option performs the first procedure and the -group(#)- option performs the second. The # in the -at- option specifies the exact value(s) at which the variable should be broken, while in the -group- option the # indicates how many, close to equal, groups the new variable should contain. If you decided that you wanted the bmi variable to be condensed into a 10-category ordinal variable, you would need to invoke the -group(10)- option.

Type egen ordbmi = cut(bmi), group(10) into the Command window and press Enter. Then type tab ordbmi into the Command window and press Enter. The new distribution of the ordinal version of the BMI of respondents is displayed as follows:

```
     ordbmi |      Freq.     Percent        Cum.
------------+-----------------------------------
          0 |        246        9.80        9.80
          1 |        254       10.12       19.93
          2 |        252       10.04       29.97
          3 |        246        9.80       39.78
          4 |        244        9.72       49.50
          5 |        263       10.48       59.98
          6 |        251       10.00       69.99
          7 |        251       10.00       79.99
          8 |        249        9.92       89.92
          9 |        253       10.08      100.00
------------+-----------------------------------
      Total |      2,509      100.00
```

Now instead of being an interval-ratio variable, ordbmi is a 10-category ordinal variable. Notice that the groups have been broken almost equally to contain 10% of the respondents. This variable could now be used to examine a cross-tabulation with gender, for example.

A second group of commonly used -egen- functions are referred to as "row" functions. A row function performs an operation on each case (i.e., a row in the data set). Usually, the argument portion of row functions involves a list of variables. The function then performs the operation based on each case's values on the specified variables.

For instance, in Chapter 7 you used respondents' perceived ideal age of marriage as a predictor of their work hours. You may have noticed that you used the variable marrymin, which only contained the minimum age for respondents that reported a range (e.g., 23–25). Instead of only using the minimum age for these cases you might consider using the midpoint (i.e., average or mean) of the two values provided in the range.

If the respondents reported a range, the maximum value is placed in the variable marrymax, but if they did not give a range, they are set as .s (i.e., skipped) on the marrymax variable. The -gen- command could be used to create the average using an expression such as marrymin + marrymax/ 2. The missing pattern, however, would cause some problems for this command. You would have to use a -replace- command to fill in the values for respondents who provided only one age because Stata cannot complete an addition when one of the values is missing. A quicker way to create this variable is to use the -egen- command with the -rowmean- function. The -rowmean- function, as do most row functions, ignores missing values. So when you calculate the row mean of the marrymin and marrymax variables for a case that is missing on marrymax, Stata will automatically calculate this case's average only using the marrymin variable. In this case, because only two variables are specified, that means Stata would divide the case's marrymin value by 1, thereby placing the single age given into the new variable for these cases. If a case was missing on both values, however, it would be set as missing on the new variable too.

Type egen idlmarry = rowmean(marrymin marrymax) in the Command window and press **Enter**.

A brief look at the data clearly shows how this command worked. Type browse marrymin marrymax idlmarry into the Command window and press **Enter** to bring up the Data Browser window for these variables.

	marrymin	marrymax	idlmarry
22	28	.s	28
23	23	.s	23
24	27	.s	27
25	20	25	22.5
26	26	.s	26
27	25	.s	25
28	28	.s	28
29	25	26	25.5
30	25	.s	25
31	26	.s	26

The command worked as intended. The first three cases shown in this window were all skipped from the marrymax variable because they provided an exact ideal age. They are then given that exact age on the new idlmarry variable. The fourth case in this window (the 25th case overall) provided a range of 20 to 25 as the ideal age of marriage. This case is then assigned the average of 22.5 (20 + 25/2) on the new idlmarry variable.

mark and markout

In Chapter 7, the issue of listwise deletion of missing cases was covered. Listwise deletion refers to the analytic strategy of conducting analyses only on cases that have valid responses on *all* the variables included in a particular statistical test. For example, the final regression analysis performed in Chapter 7 only used 2,455 of the 2,532 cases in the data set. There were 77 cases that had missing responses to one or more variables included in the analysis.

Stata typically performs a listwise deletion by default within any one particular command. But when conducting a research project it is typically advisable to always use the same set of cases for all the analysis performed. Following the example from Chapter 7, when presenting the descriptive statistics for the variables involved in the regression analysis, it is more appropriate to use only the 2,455 cases that are present in the final analysis rather than the total valid cases for each particular variable.

It would be possible, after selecting all the variables that will be included in the final analysis, to always perform a detailed -if- statement for all the commands used in the production of results. For example, four variables were used in the final analysis in Chapter 7: workhr, totacts, marrymin, and currdate. (The totacts and currdate variables were created in Chapter 7 but are already included in the Chapter 8 Data.dta file). The initial -sum- output of the workhr variable shows that 2,527 cases were used to produce the given measures of central tendency and variability, meaning the workhr variable only has 5 missing cases. To limit those calculations to only the final valid sample of 2,455 cases, type sum workhr if totacts<. & marrymin<. & currdate<. into the Command window and press **Enter**.

```
    Variable |       Obs        Mean    Std. Dev.       Min        Max
-------------+--------------------------------------------------------
    workhrs1 |      2455    20.42851    18.92613          0        100
```

This -if- statement has told Stata to conduct the -sum- command on cases that are less than missing on all the variables included in the -if- clause. Remember that Stata treats all missing coded responses as the highest values on the variable, so specifying <. in the -if- statements inherently directs Stata to use cases with valid responses on the variables.

This method for limiting all the analyses in a project to the same set of cases is effective but can become very cumbersome if numerous variables are included in the final analyses. Even more problematically, if a variable is added or removed from the analyses, every single -if- statement would have to be changed. Clearly, this possibility emphasizes why using a do file can be so helpful. But even if you use a do file, changing all the necessary -if- statements would be time-consuming.

Stata provides a two-command procedure that provides an easier way to identify and use the final analytic sample cases. The basic structure of the two commands is as follows:

```
mark newvar
markout markvar varlist
```

The first command, -mark-, creates a variable (i.e., newvar) that will be used to identify cases, coded 1, that have valid responses on all the variables to be included in the analysis. You can name this variable anything that you would like, and after executing this command you will see the new variable appear in the Variables window. At this point, however, every single case in the data set is coded as 1.

The second command -markout- is where you specify the variables that are to be included in the final analysis. The -markout- command then automatically recodes the variable created in the -mark- command to 0 if a case is missing on *any* of the variables listed. It may sound complicated, but following a step-by-step example should help clarify matters.

Type mark nomiss into the Command window and press **Enter**. This command creates a variable, called nomiss, that will identify the nonmissing cases.

Next, type markout nomiss workhrs totacts marrymin currdate in the Command window and press **Enter**. Producing a frequency distribution of the new variable illustrates what has been accomplished (-tab nomiss-):

```
     nomiss |      Freq.     Percent        Cum.
------------+-----------------------------------
          0 |         77        3.04        3.04
          1 |      2,455       96.96      100.00
------------+-----------------------------------
      Total |      2,532      100.00
```

The nomiss variable is now an indicator of the 2,455 cases that have valid data on the four variables that are included in the final regression analysis.

Now, you can type the previous lengthy -sum- command as: sum workhr if nomiss==1, and the appropriate results will be presented.

At first, it may seem to be a hassle to take the time to create this marked variable. But as the number of analyses that are included in your research project increases, it can save a significant amount of time and effort. This procedure, however, is best suited to analyses that are conducted with do files. If you are using do files, you can write all of the commands to include the -if nomiss==1- clause. Then if you decide to add or delete a variable from the analyses, you can drop the current nomiss variable, modify the -markout- command accordingly, recreate the corrected nomiss variable, and then quickly rerun the analyses on the newly identified sample.

alpha, gen(newvar)

The final convenience command actually performs two useful functions. First, it displays a commonly used statistic in analysis: Cronbach's alpha. Second, with the use of an option, it can create an index variable based on the set of variables included in Cronbach's alpha calculation.

Before giving away the structure of this command, this is a useful example to illustrate the value and practice of using Stata help files. If you are thinking about how you would tell a smart colleague to produce Cronbach's alpha, you might consider "cronbach." If you type help cronbach into the Command window and press Enter, you notice your first intuition was not quite correct. But in the search results window you see that the appropriate command appears as the first result: -alpha-. Now you can either click on "alpha" in the results window or type help alpha in the Command window, and press Enter, to see the full help file associated with producing Cronbach's alpha.

Cronbach's alpha (from here on simply alpha) is a statistic that assesses how well a set of variables "hang together." It is typically used to determine whether a group of variables can be validly combined into a single index or scale. It can range from 0 to 1, with values close to 1 indicating the items are closely related and form an effective index.

For example, the NSYR contains three questions that seem to measure the underlying concept of self-worth. The questions asked, "In general how often do you (1) feel loved and accepted for who you are; (2) feel alone and misunderstood; (3) feel invisible because people don't pay attention to you?" Each question has four response options ranging from "None" to "A Lot." The variables' names are accepted, alienate, and invisibl.

To calculate the alpha value of these variables, type `alpha accepted alienate invisibl` into the Command window and press **Enter**.

```
Test scale = mean(unstandardized items)
Reversed item:  accepted

Average interitem covariance:      .2062425
Number of items in the scale:             3
Scale reliability coefficient:       0.6252
```

The results display the average interitem covariance, the number of items in the scale, and the scale reliability coefficient. This last figure is Cronbach's alpha. In the example, the value of .6252 suggests that the three variables are moderately related.

The display also notes that the variable `accepted` has been "reversed." Stata automatically alters the direction of the variables so that they are all indicating the underlying measure. The three variables in the example are being used as indicators of self-worth. High values on the `accepted` variable (i.e., respondents who feel accepted "a lot") would indicate high self-worth, whereas high values on the `alienate` and `invisible` variables would indicate a low self-worth (i.e., the respondent feels alienated and invisible "a lot"). Therefore, either the `accepted` or both the `alienated` and `invisible` variables need to be reversed so that similar values indicate the same aspect of the underlying measure.

This automatic reversal is particular useful when the `-gen(newvar)-` option is invoked. This option instructs Stata to generate a new variable that is the average value of all the variables included in the `-alpha-` command. When all the values of the variables are not aligned in a similar direction, as in the current example, Stata automatically reverses the values of the necessary variable(s) before calculating the average. In doing so, Stata ensures that the resulting index is created correctly.

Type `alpha accepted alienate invisibl, gen(worth)` into the Command window and press **Enter**. The same results from above are presented again, and now a new variable, called `worth`, is shown in the Variables window.

One quick note about how Stata generates the new variable in the `-alpha, gen(newvar)-` command. Every case that has a valid response for at least one of the included variables is assigned a value on the new variable. The average is based on the number of valid responses each case has. For example, if a case was missing on the `alienate` variable, his or her value on the newly generated `worth` variable would be calculated as the value on `accepted` plus `invisible` divided by 2 (not 3). This information is provided in the help file. There you also see that you can specify the -casewise- option if you prefer that cases with missing responses on any of the included variables be set to missing on the generated variable.

Summary of Commands Used in This Chapter

```
*tab,gen()
tab denom, gen(dumden)
tab dumden2

*egen
egen ordbmi = cut(bmi), group(10)
tab ordbmi
egen idlmarry = rowmean(marrymin marrymax)
browse marrymin marrymax idlmarry

*mark markout
sum workhrs if totacts<. & marrymin<. & currdate<.
mark nomiss
markout nomiss workhrs totacts marrymin currdate

*alpha
alpha accepted alienate invisibl
alpha accepted alienate invisibl, gen(worth)

*A Closer Look: Using Wildcards
tab1 dumden*
tab1 dumden1 - dumden6
```

Exercises

Use the original `Chapter 8 Data.dta` for the following problems. [Optional: Complete the exercises by using a do file and save the results using a log file. See Chapter 3 for an explanation of how to use these files.]

1. Create a set of dichotomous indicator variables for each possible view of God (`godview`).

2. Display a frequency distribution of each indicator variable to ensure the dummy set was created correctly. [Bonus: Try using wildcards.]

3. Use the Command window to examine the Help File for the `-egen-` command.

4. Based on the `-egen-` Help File, execute a command that generates a new variable containing the minimum value of number of people dated (`datnum`) and number of friends (`numfrien`).

5. Display the measures of central tendency for the `workhrs1`, `datnum`, and `numfrien` variables that includes only cases with valid responses on all the three variables.

6. The NSYR contains three variables about gender roles in relationship. The questions ask how much respondents agree with whether a woman can have a fully satisfying life without getting married (`wommar`), whether the man should make most of the decisions in the household (`mandecid`), and whether a working mother can have a secure and warm relationship with her children (`wrkngmom`). Use Cronbach's alpha to determine whether these three measures would form an effective scale.

7. Create an index variable measuring traditional gender roles, using the `wommar`, `mandecid`, and `wrkngmom` variables.

Appendix

Getting to Know Stata 11

For many people, learning any new computer software can be an anxiety-producing task. When that computer program involves statistics, the stress level generally increases exponentially. If you have similar feelings as you begin your journey into becoming a Stata user, do not fear, you are not alone. This book is designed with this apprehension in mind. One of the primary goals of this book is to help alleviate, or at least minimize, this anxiety as we move toward becoming an effective and proficient Stata user. Keep in mind that at one time you may have had similar feelings about using e-mail or the Internet, and just as many people now feel extremely comfortable using these programs. By the end of this book you will have a similar grasp of and comfort with Stata.

Before diving into all the details of using Stata, it is important to have an understanding of its various components. This chapter will serve as an introduction to the basic building blocks of the Stata program. Each of these aspects will be covered in much more detail throughout the book, but this chapter provides an overview of the basic functionality of the Stata program. The second section of the chapter explains how data are opened, imported, and entered.

What You See

When you open Stata, by double clicking on the Stata icon, for the first time, you will see the following screen:

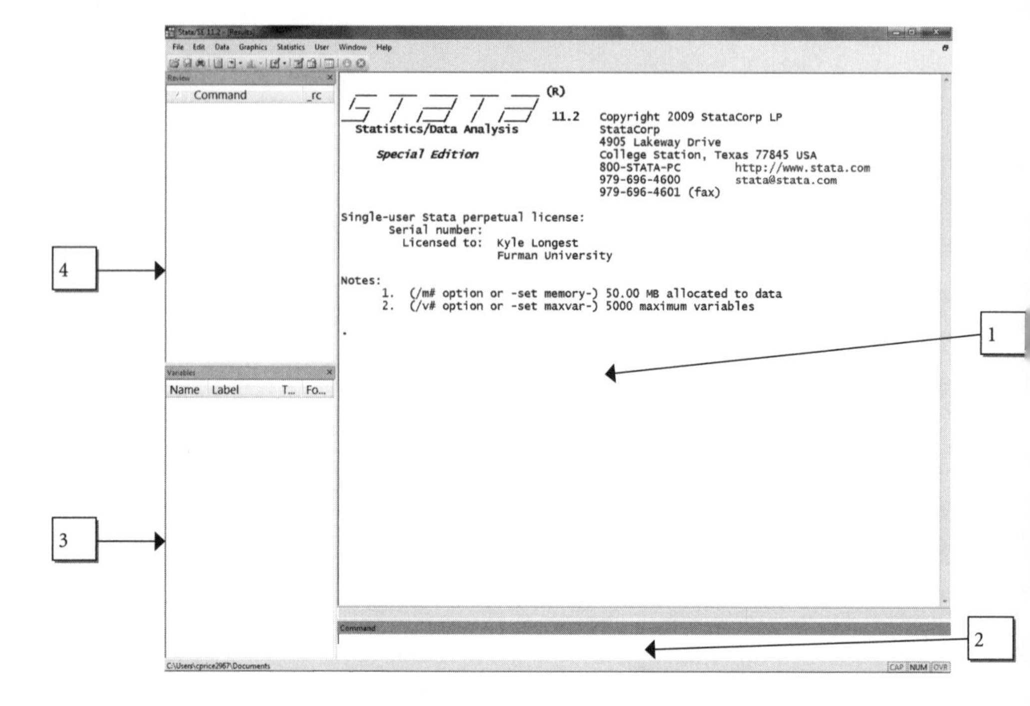

There are four different windows on the screen.[1]

1. Results Window: The Results window is where everything that Stata "does" will be displayed. Anytime Stata executes some operation, it will display that operation and its results in this window. These results, however, are not automatically saved. How to save these results is covered in the Data Management: Saving Results section of Chapter 3.

2. Command Window: The Command window is where you will enter the operations that you want Stata to perform, when using the "syntax" interface. Syntax, or code, is another term for Stata's command language. These are the words that tell Stata what procedures to execute. Commands are entered, one at a time, in this window. After you type a command into the Command window, pressing the Enter key on your keyboard makes Stata execute the procedure

[1]This layout is what you would see if Stata was opened "right out of the box." If you are working on a shared computer (or over a network), there is a chance that these windows have been moved, resized, or even deleted by another user, making what you see slightly different from the screenshot presented. If any of these windows are missing, you can click on the Windows tab and click on the desired window. You also can move these windows by simply clicking on them with your mouse and dragging them to the desired location.

that is defined by the Command. One helpful feature of the Command window is that you can scroll through previously executed commands by pressing the Page Up key. When you find the previous command you are interested in, you can either alter it or simply press Enter again to rerun the same command. The majority of this book will be devoted to explaining and describing the various commands that you will need to use to perform quantitative analyses.

3. Variables Window: When you open a data file in Stata, the variables contained in that data set will be listed in the Variables window. This window can be used to scroll through and see all the variables that are contained in the active data. Whenever you click on a variable name listed in the Variables window, it will automatically appear in the Command window. This window also lists the variable "Label," which presents more detailed information about the variable. Labels are discussed in more detail in the Data Management: Working With Labels section of Chapter 3.

4. Review Window: The Review window contains a running history of all the operations that have been performed during the current session of Stata. Whenever you enter and execute a command, it will appear both in the Results window and in the Review window. The most useful aspect of the Review window is that it can be used as a shortcut to work with a previously executed command. When you click on a command in the Review window that command will appear in the Command window, from which you can alter the command or simply rerun the same command.

There also are several icons at the top of the screen. The purpose and use of these icons are covered throughout the book. Each of these basic windows will become familiar to you as we go through this book. For now, be sure that you feel comfortable identifying the main purpose of each of the windows.

Getting Started With Data Files

When working with Stata, you will be using what is referred to as a "data file." If you are familiar with typical database programs, then you already know what a data file basically is. These files contain information (often numerical) on a set of cases, such as respondents to a survey, a sample of schools, or each of the states in the United States. Generally, data files are organized such that information regarding each case is contained in one row in the file, whereas each column represents a variable (i.e., information about that case), such as a person's gender, a school's total number of students, or a state's total square miles.

Similar to most computer files, data files come in many different types. But just like a PDF file is very similar to a word document, so too are all data files essential derivations of a similar structure. Each of these derivations is denoted by a different file extension—the letters that come after the "." in a file name. The primary file for Stata data files is .dta. Moving other types of data files into Stata (e.g., Microsoft Excel files) is covered in the Using Different Types of Data Files in Stata section of this chapter.

OPENING AND SAVING STATA DATA FILES

To open a data file that is in Stata format (i.e., one that has a .dta extension), select the File menu (in the upper left-hand corner), then choose Open. Or alternatively, you can simply click on the icon. From here you will need to search through the disk drives and folders on your computer to find your saved data file. This chapter uses the data file, available at **www.sagepub.com/longest**. named `Chapter 1 Data.dta`. Once you have found your data file, double click the file. Having done this, you will notice that the Stata screen looks different from how it did initially.

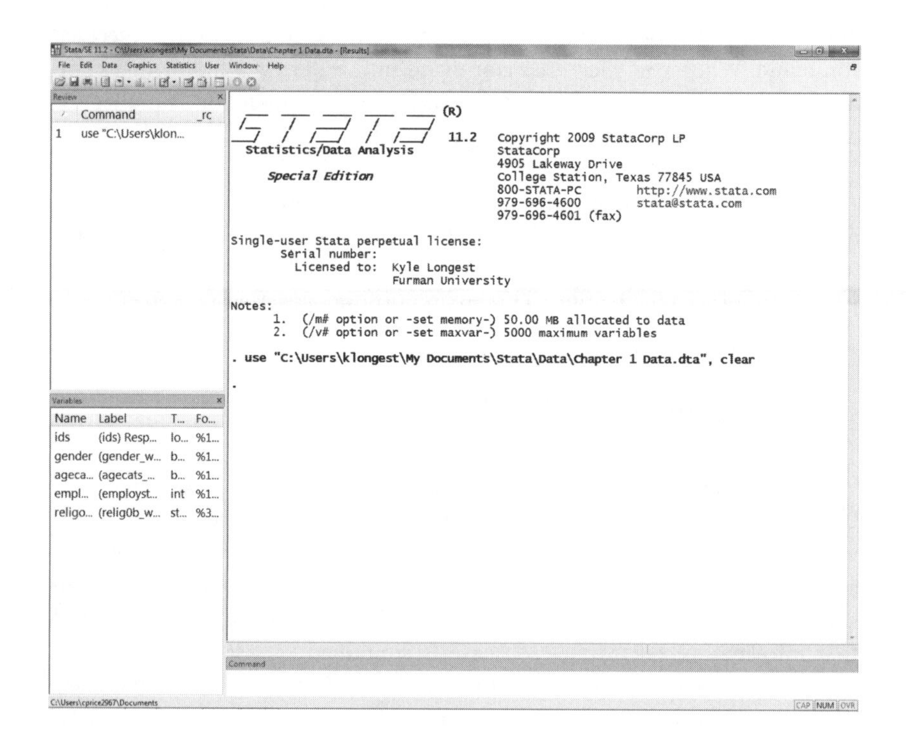

The first operation you performed is now displayed in both the Results and Review windows. Again, whenever we tell Stata to "do" something, whether through the point-and-click menus or by entering a command in the Command window, it will be displayed in the Results and Review windows. Because opening a data file does not have any "results," only the command is displayed in the Results window. You also can see that the data file contains five variables, listed in the Variables window. All the information provided about each variable in this window is discussed in a later section, but for now the most important aspect is the variable name. In this data set, the five variables are named `ids`, `gender`, `agecats`, `employst`, and `religoth`. These variable names should give you some indication of what type of information the variable contains. The variable gender, for example, says whether each respondent is a male or a female.

It is a good practice to always save a copy of your data files and only work with that duplicated version. When working with and analyzing data, you will often be forced to change aspects of the data files. For example, you may need to create a new variable or change something about an existing variable. But it is important to have an original version of the data, just in case something undesired occurs. Don't worry too much; most alterations you perform can be undone or recovered. Working with a duplicate copy of the data is simply an added protection.

To save a duplicate copy of the data file you have just opened, open the **File** menu and click on **Save As**. You can then enter a new file name, such as `Chapter 1 Data mycopy.dta`, and click **Save**. This is the procedure you will use whenever you want to save a new version of your data file.

A Closer Look: Stata Data Files Across Versions

As was noted in the Preface, the vast majority of Stata features and commands are similar across versions (e.g., Stata 12, 11, 10, etc.). This is true of Stata data files, by and large. All Stata data files that are created and/or saved in an older version can be read by a newer version (i.e., forward compatible). That means that if you are using Stata 11 but are working with colleagues who are using Stata 12, any files that you send to them will open without a problem.

(Continued)

(Continued)

During certain upgrades, however, Stata data files cease being "backward" compatible, meaning files saved in a newer version cannot be opened by older versions. Stata 12 happens to be one of those upgrades. If your colleagues are using Stata 12 and send you a data set that was saved in Stata 12, you will not be able to open it in Stata 11. [Note: This is not a problem if you are moving files between Stata 11 and Stata 10, as these two versions are completely compatible with each other.]

Do not despair. Stata has built in a very simple feature to overcome this problem. If you know that you want the data you (or your colleagues) are using in Stata 12 to be opened by older versions, you (or they) need to take one extra step (from the process just explained). [Also, note that if you are using Stata 11 and are working with a colleague using a version prior to Stata 10, you would need to follow these same steps.]

First, click on the **File** menu, then click on **Save As**. Now, use the drop-down menu in the **Save as Type** box and select **Stata 9/10 Data (*.dta)** option (or **Stata 9 Data (*.dta)** if you are using Stata 11). The option is listed as "Stata 9/10" and not 11 because Stata Versions 8 and 9 as well as 10 and 11 are completely compatible with each other (both forward and backward), so using this option actually allows the data to be opened in any version of Stata from 8 through 12. Note that you do not need to change the file extension, it is still `.dta`. Once you have named your file, click **Save**. You will know that you have saved the data correctly when the output in the Results starts with `.saveold`, which is telling you that the file has been saved in a way that makes it readable by the previous versions. Again, note that when you save a file in this way, it can still be used in Stata 12 (or any newer version).

DATA BROWSER AND EDITOR

If this is the first time you are working with data, it may be helpful to actually "see" the data. Even if you have experience using data, it may often be helpful to look at the data you are examining. To see the data file in Stata, you can click on the Data Browser icon, , in the middle of the top of the screen. When you do so, you will see a new window that appears as shown below:

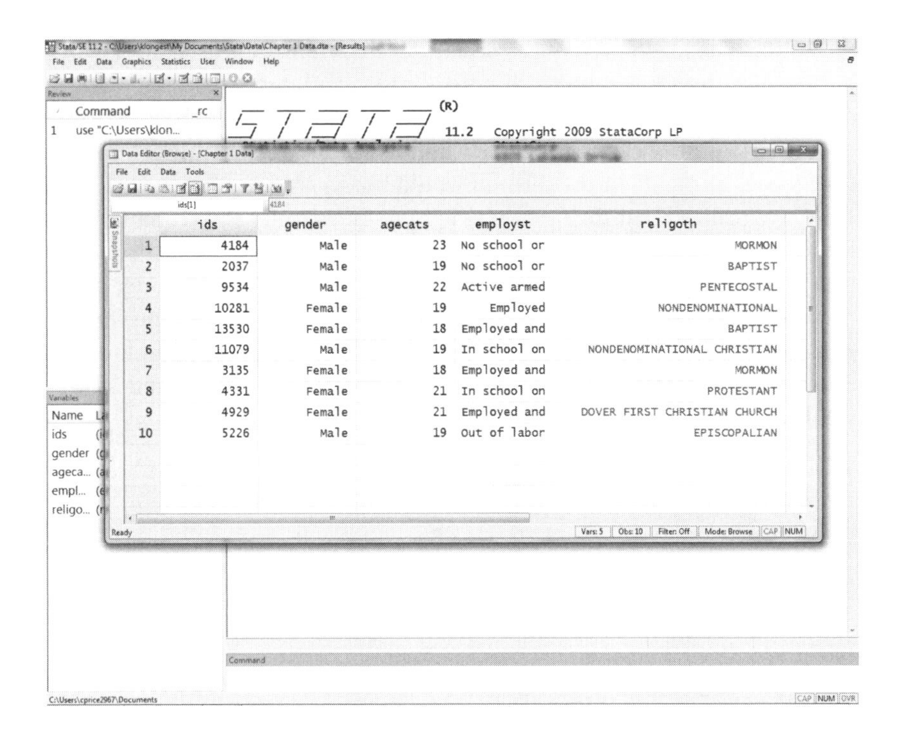

This new window, as is denoted in its upper left-hand corner, is the Data Editor (Browse) window. The "(Browse)" aspect indicates that you are only looking at the data, not actually changing them.

In this window, you see all five of the variables that were listed in the Variables window. As was mentioned earlier, each row is a different case (i.e., a National Study of Youth and Religion [NSYR] respondent), and each column is a different variable. Each cell then contains information on the given variable for that case. For example, the case in the first row is a "Male" respondent who mentioned that "Mormon" was his other religion. To close this window, click on the red "X" in the upper right-hand corner.

There may be times when you want to change the value of a particular case on an individual variable. One way to do so is by using the Data Editor window. (A more efficient way to change the values of multiple cases is covered in The 5 Essential Commands: replace (if) section of Chapter 2.) To begin, click on the Data Editor icon, , which is next to the Data Browser icon. You may notice that the Data Editor and Data Browser windows look very similar. The main difference is that in the upper left-hand corner of the window, after "Data Editor," the window now reads "(Edit)." It is important to know which window you have opened because you can change the values of the data when the Editor is open. To prevent any accidental alterations, it is generally advised only to use the Data Browser window unless you are certain you want to change a particular value.

After you have opened the Data Editor window, use the direction keys (or mouse) to highlight the cell you would like to change. For example, you may have realized that the first case's age was incorrectly entered in the data file. Instead of being 23 years old, this case should only be 22 years old. To make this change, once you have the cell in the first row listed under `agecats` highlighted, type 22 and press **Enter**. This case's value for the variable `agecats` has now changed. When you close the Data Editor window, this operation has been recorded and displayed in both the Review and Results windows.

A Closer Look: Your First Command

You may have noticed that when you changed the first case's value using the Data Editor window, the following text was displayed in the Results window:

```
. replace agecats = 22 in 1
(1 real change made)
```

Whenever you use the menus or a point-and-click method for performing an operation in Stata, it displays the command that would be entered in the Command window to perform the same operation in the Results window. In this Data Editor example, you can see that the command to change a value is `-replace-`. If you had entered this full command into the Command window and pressed **Enter**, the same change would have been made. At times, it may be helpful to perform an operation for the first time using the menus, but, as discussed in much more detail in Chapter 2, it is extremely beneficial to know and use the commands via the Command window for the majority of the operations you need to perform.

The rest of this book will discuss how to perform operations using the Command window. But to see the connection between the menu-based operation and the Command window, try this: Type (or copy and paste) the full command (except the first ".") that was displayed in the Results window when you closed the Data Editor window into the Command window. Now change the "22" to "23." The command should read

```
replace agecats = 23 in 1
```

Then press **Enter**. Open the Data Browser window again and notice the change to the first case's value under `agecats`.

ENTERING YOUR OWN DATA

Many data files that you will analyze will already be in Stata format or in a format that can be easily converted to Stata format (more on this topic below). Yet there may be times when you need to enter the data from a study. For example, if you distributed a survey through the mail, you will need to input the responses to each question for each case so that you can analyze them in Stata.

The first step in entering your own data after you have opened Stata is to open the Data Editor window as above. From here you can simply enter the values for each case on each variable. Entering data in this way is very similar to entering values into a Microsoft Excel file. The Data Editor, however, does not have the equation functionalities that an Excel file would.

When you begin entering values, each variable is automatically named `var1`, `var2`, and so on. Most often it is helpful to have the variable names be more descriptive of the values they contain. One way to change these generic names to something that more clearly identifies the variable is to right click on the current name of the given variable you want to rename (e.g., `var1`) listed at the top of the Editor window. Then click on **Variable Properties** and fill in the desired name in the **Name** blank. Another option would be to close the Data Editor window when you have finished entering all of the data. Then you can right click on the variable name (e.g., `var2`) in the Variables window, followed by clicking on the **Rename 'var2'** option. Again, simply type the new name in the blank.

Once you have finished entering all of your data, close the Data Editor and follow the steps described above to save a copy of your data file in Stata format.

USING DIFFERENT TYPES OF DATA FILES IN STATA

Some data files may not be available in Stata format. Therefore, a few steps are needed to work with these files in Stata. It would be virtually impossible to cover every possible data file type and how each can be transferred to be usable in Stata. Instead, the most common type will be covered. Also note that there are other computer software programs that are specifically designed to convert data files into various formats (e.g., Stat/Transfer). If you have access to such a program, it is probably the most effective and efficient way to transfer files into a Stata format. Some statistical software packages also offer the option of saving a data file in a different format, which often includes the Stata, .dta extension.

One of the most frequently encountered data file type that is not Stata-ready is a Microsoft Excel file. Usually these files are denoted with the .xls extension, but other extensions (e.g., .csv) that are generated or readable by Microsoft Excel can all be treated in a similar fashion.

This process requires that you have access to and some familiarity with Microsoft Excel. To start, open the data file in Microsoft Excel. Then highlight the entire worksheet that contains the data and copy it (either by right clicking and choosing Copy or using the copy function (Ctrl+C)). Next, in Stata open the Data Editor window, highlight the upper left data cell, right click and choose Paste, or use the paste function (Ctrl+V). Once you pasted in the data, you should be presented with a window that asks whether you want to Treat First Row as Data or Treat First Row as Variable Names. The option that you choose will depend on whether your Excel file contains variable names in the first row or whether it contains only data. The two formats are shown below.

First Row as Variable Names

	A	B	C	D	E
1	ids	gender	agecats	employst	religoth
2	4184	Male	23	No school or	MORMON
3	2037	Male	19	No school or	BAPTIST
4	9534	Male	22	Active armed	PENTECOSTAL
5	10281	Female	19	Employed	NONDENOMINATIONAL
6	13530	Female	18	Employed and	BAPTIST
7	11079	Male	19	In school on	NONDENOMINATIONAL CHRISTIAN
8	3135	Female	18	Employed and	MORMON
9	4331	Female	21	In school on	PROTESTANT
10	4929	Female	21	Employed and	DOVER FIRST CHRISTIAN CHURCH
11	5226	Male	19	Out of labor	EPISCOPALIAN
12					
13					
14					
15					
16					
17					
18					
19					

First Row as Data

	A	B	C	D	E	F
1	4184	Male	23	No school or	MORMON	
2	2037	Male	19	No school or	BAPTIST	
3	9534	Male	22	Active armed	PENTECOSTAL	
4	10281	Female	19	Employed	NONDENOMINATIONAL	
5	13530	Female	18	Employed and	BAPTIST	
6	11079	Male	19	In school on	NONDENOMINATIONAL CHRISTIAN	
7	3135	Female	18	Employed and	MORMON	
8	4331	Female	21	In school on	PROTESTANT	
9	4929	Female	21	Employed and	DOVER FIRST CHRISTIAN CHURCH	
10	5226	Male	19	Out of labor	EPISCOPALIAN	
11						
12						
13						
14						
15						
16						
17						
18						
19						

After you have selected the option that fits with the type of data file you have, close the Data Editor, and follow the previously described steps to save the data from within Stata as a Stata data file. Once you have saved your data as a Stata data file, you can simply open and use this version of your data.[2]

TYPES OF VARIABLES IN DATA FILES

At this point, you should feel comfortable with the basic structure of data files. Each row holds the information for one case and each column is a different variable. With this knowledge, you are almost ready to start analyzing your data. There is, however, one distinction in the types of variables included in data that is important to understand.

[2]This "copy and paste" method is the easiest way to transfer data from Microsoft Excel into a Stata format, especially for novice users. But there are some disadvantages to this strategy. More practiced users should transform Excel worksheets into .csv files and then implement the `-insheet-` command. The specifics of this command are beyond the scope of this introductory text, but the Stata Help Files section of Chapter 8 provides information on how Stata's Help files can be used to learn how to use this command.

To help illustrate this difference, consider the NSYR variable, in the `Chapter 1 Data.dta` file, `gender`. This variable came from the following question asked of all respondents:

Are you

a. Male?

b. Female?

If you were entering the responses to this question into a Stata data set, you could record them in one of two ways. First, the actual answer "Male" or "Female" could be recorded for each case. Second, you could use a number to represent each answer. For example, you could choose to enter 0 for all respondents reporting "Male" and 1 for all respondents reporting "Female."

If you record the responses in the first way, it would be what Stata refers to as a *string* variable. A string variable is a variable in which the contents are actual words. String variables can be very useful for many purposes. For example, you can enter verbatim answers to questions directly into Stata, as was done for the variable `religoth` in the `Chapter 1 Data.dta`.

The drawback of storing a variable such as gender as a string variable is that some statistical operations require numbers. For example, if you wanted to calculate the mean (i.e., mathematical average) of a variable, each category must be assigned a numeric value. For this reason, it is generally advisable, when possible, to use the second method and enter variables as *numeric* variables. These are variables that have actual numbers attached to each response.

Fortunately, many of the Stata commands that will be discussed in this book operate similarly with numeric or string variables. The commands that work only with numeric variables are those that perform statistical operations that require numbers to calculate, for example, the mean or a linear regression. Because numeric variables, typically, are more applicable to the vast majority of data analyses, the commands discussed in this book focus on their use with numeric variables (keeping in mind that many operate identically for string variables). The primary commands that are used (and are different) for string variables, including methods for changing a string variable to a numeric variable are addressed in the Data Management: Using String Variables section in Chapter 3.

As has been discussed, often you may be using data that you did not enter, so you may not have a choice or even be certain about the way in which variables were entered. There are several ways to determine whether a variable is a numeric or string variable. The most straightforward way is to open the Data Browser window. In versions Stata 10 or later, string variables are shown in a red font, whereas numeric variables are shown in either black or blue font. In the `Chapter 1 Data.dta` file, you will see that only the variable `religoth` is a string variable.

A Closer Look: Variable Types

You may have noticed that more information about the variable type is listed in the Variables window. For example, `gender` is shown to be a byte variable, `ids` is a long variable, and `religoth` is a str31 variable.

These distinctions further demarcate variables within the general categories of numeric and string. They also are related to how much file space is allotted to storing the variable.

All string variables have the "str" prefix, and the number indicates the maximum characters that can be used for that string variable. So the maximum length a denomination could be in the variable `religoth` is 31 characters. As you will see, this constraint can be altered, but it is advisable to use only the minimum number of characters that are needed. Otherwise you are using memory to store empty spaces.

Similarly, the various subtypes of numeric variables indicate the number of digits that each variable can hold. In order of smallest to largest, the numeric variable types are byte, int, long, float, and double.

Generally, Stata will store variables in the most efficient and effective way when you create them. Moreover, most users of Stata will conduct countless analyses without ever having to worry or manipulate these specific distinctions.

When you have the Data Browser open, you probably notice, however, that the variables `gender` and `employst` look different from the variables `ids` and `agecats`. This difference is due to the fact that `gender` and `employst` have what are called *value labels* attached to them. Value labels will be covered in much more detail later, but they are labels that can be applied to the numeric codes used to represent responses. Remember that you could decide to use the number 1 to represent the answer "Female." This choice may be difficult to remember (i.e., whether 1 was Male or whether

1 was Female), therefore you can use value labels as a shortcut to help remember this coding strategy. The variables `ids` and `agecats` were numerical responses so they do not have any value labels that could be attached to them. You can see the actual numerical codes for each variable using the Data Browser window by clicking on the **Tools** menu, selecting **Value Labels**, and clicking **Hide All Value Labels**. When you do so you will see the cases that were "Male" now display "0" and the cases that were "Female" now display "1." Or you can highlight (either using the direction keys or the mouse) a particular cell (e.g., "Male"). When you do so, the actual value is listed in a pane just underneath the icons.

Chapter Exercise Solutions[1]

Chapter 1 Exercises – Solutions

3. 10 cases, 6 variables

4. `cendiv` is a string variable, the rest are numeric.

6.

The Data Editor should appear as below:

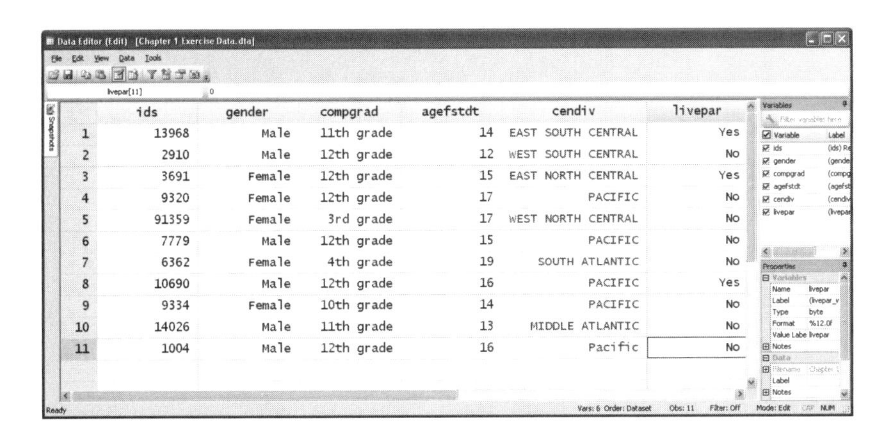

	ids	gender	compgrad	agefstdt	cendiv	livepar
1	13968	Male	11th grade	14	EAST SOUTH CENTRAL	Yes
2	2910	Male	12th grade	12	WEST SOUTH CENTRAL	No
3	3691	Female	12th grade	15	EAST NORTH CENTRAL	Yes
4	9320	Female	12th grade	17	PACIFIC	No
5	91359	Female	3rd grade	17	WEST NORTH CENTRAL	No
6	7779	Male	12th grade	15	PACIFIC	No
7	6362	Female	4th grade	19	SOUTH ATLANTIC	No
8	10690	Male	12th grade	16	PACIFIC	Yes
9	9334	Female	10th grade	14	PACIFIC	No
10	14026	Male	11th grade	13	MIDDLE ATLANTIC	No
11	1004	Male	12th grade	16	Pacific	No

[1]The commands to produce the correct solutions are shown in bold font. Solutions are only presented for questions that have "visible" responses. For example, the first question in the Chapter 1 Exercises asks you to open a data file. There is no answer that could be shown for this question, therefore it is not listed.

Chapter 2 Exercises – Solutions

1.

tab numfrien

```
  (numfriend_w3) |
     N:1. Now for |
 the next set of |
 questions, I'll |
  be asking some |
          things |      Freq.     Percent        Cum.
-----------------+-----------------------------------
               2 |          2        8.00        8.00
               3 |          8       32.00       40.00
               4 |          5       20.00       60.00
               5 |         10       40.00      100.00
-----------------+-----------------------------------
           Total |         25      100.00
```

2.

tab numfrien, sort

```
  (numfriend_w3) |
     N:1. Now for |
 the next set of |
 questions, I'll |
  be asking some |
          things |      Freq.     Percent        Cum.
-----------------+-----------------------------------
               5 |         10       40.00       40.00
               3 |          8       32.00       72.00
               4 |          5       20.00       92.00
               2 |          2        8.00      100.00
-----------------+-----------------------------------
           Total |         25      100.00
```

3.

tab numfrien agecats

```
  (numfriend_w3) |
     N:1. Now for |
 the next set of |
 questions, I'll |
  be asking some |   (agecats_w3) Age variable collapsed into one year categories
          things |        18         19         20         21         22         23 |     Total
-----------------+------------------------------------------------------------------+----------
               2 |         0          0          1          0          0          1 |         2
               3 |         1          1          2          2          2          0 |         8
               4 |         2          0          2          0          1          0 |         5
               5 |         4          1          1          2          2          0 |        10
-----------------+------------------------------------------------------------------+----------
           Total |         7          2          6          4          5          1 |        25
```

4.

tab numfrien agecats, col

```
+-------------------+
| Key               |
|-------------------|
|     frequency     |
| column percentage |
+-------------------+

 (numfriend_w3) |
  N:1. Now for  |
the next set of |
questions, I'll |
be asking some  | (agecats_w3) Age variable  collapsed into one year  categories
        things  |     18        19        20        21        22        23 |   Total
----------------+------------------------------------------------------------+-------
             2  |      0         0         1         0         0         1 |       2
                |   0.00      0.00     16.67      0.00      0.00    100.00 |    8.00
----------------+------------------------------------------------------------+-------
             3  |      1         1         2         2         2         0 |       8
                |  14.29     50.00     33.33     50.00     40.00      0.00 |   32.00
----------------+------------------------------------------------------------+-------
             4  |      2         0         2         0         1         0 |       5
                |  28.57      0.00     33.33      0.00     20.00      0.00 |   20.00
----------------+------------------------------------------------------------+-------
             5  |      4         1         1         2         2         0 |      10
                |  57.14     50.00     16.67     50.00     40.00      0.00 |   40.00
----------------+------------------------------------------------------------+-------
         Total  |      7         2         6         4         5         1 |      25
                | 100.00    100.00    100.00    100.00    100.00    100.00 |  100.00
```

5.

gen totties=numfrien+datnum

tab totties

```
    totties |      Freq.     Percent        Cum.
------------+-----------------------------------
          3 |          1        4.00        4.00
          4 |          1        4.00        8.00
          6 |          3       12.00       20.00
          7 |          3       12.00       32.00
          8 |          5       20.00       52.00
         11 |          2        8.00       60.00
         12 |          1        4.00       64.00
         13 |          1        4.00       68.00
         14 |          1        4.00       72.00
         15 |          1        4.00       76.00
         16 |          1        4.00       80.00
         19 |          2        8.00       88.00
         23 |          2        8.00       96.00
         25 |          1        4.00      100.00
------------+-----------------------------------
      Total |         25      100.00
```

6.

```
gen datnum15 = datnum
replace datnum15 = 15 if datnum==20

tab datnum15
```

datnum15	Freq.	Percent	Cum.
1	1	4.00	4.00
2	4	16.00	20.00
3	4	16.00	36.00
4	1	4.00	40.00
5	3	12.00	52.00
6	2	8.00	60.00
7	1	4.00	64.00
9	1	4.00	68.00
10	2	8.00	76.00
11	1	4.00	80.00
15	5	20.00	100.00
Total	25	100.00	

7.

```
gen olddatr = 0
replace olddatr = 1 if agecats==20 & (datnum>=5 & datnum<=10)

tab olddatr
```

olddatr	Freq.	Percent	Cum.
0	24	96.00	96.00
1	1	4.00	100.00
Total	25	100.00	

8.

```
recode agecats (18/20=0) (21/24=1), gen(age21)

tab age21
```

```
  RECODE of |
    agecats |
((agecats_w |
     3) Age |
   variable |
```

```
   collapsed |
    into one |
        year |
 categories) |      Freq.     Percent        Cum.
-------------+-----------------------------------
           0 |         15       60.00       60.00
           1 |         10       40.00      100.00
-------------+-----------------------------------
       Total |         25      100.00
```

9.

```
rename numfrien frndnum
```

10.

```
di 976*543

529968
```

Chapter 3 Exercises – Solutions

3.

```
*Chapter 3 Exercises
```

4.

```
recode datnum (0/2=1) (3/10=2) (11/100=3), gen(datlevsalt)
```

5.

```
lab var datlevsalt "Categories of People Dated"
```

6.

```
lab def datcats 1 "Minimal Dating" 2 "Moderate Dating" 3
"Extensive Dating"
lab val datlevsalt datcats
```

7.

```
tab employst
tab employst, nol

recode employst (999=.s)
```

8.

```
tab employst

    (employstat_w3) Employment |
                        Status |      Freq.     Percent        Cum.
-------------------------------+-----------------------------------
             Out of labor force |         85        3.36        3.36
 No school or work but looking |        127        5.02        8.39
                      Employed |        700       27.69       36.08
           Employed and school |        951       37.62       73.69
                In school only |        598       23.66       97.35
           Active armed forces |         67        2.65      100.00
-------------------------------+-----------------------------------
                         Total |      2,528      100.00

tab employst, mis

    (employstat_w3) Employment |
                        Status |      Freq.     Percent        Cum.
-------------------------------+-----------------------------------
             Out of labor force |         85        3.36        3.36
 No school or work but looking |        127        5.02        8.37
                      Employed |        700       27.65       36.02
           Employed and school |        951       37.56       73.58
                In school only |        598       23.62       97.20
           Active armed forces |         67        2.65       99.84
                            .s |          4        0.16      100.00
-------------------------------+-----------------------------------
                         Total |      2,532      100.00
```

9.

```
tab religoth
gen str mormonoth = "Mormon" if religoth=="MORMON" | religoth=="LATTER
DAY SAINTS" | religoth=="LATTER DAY SAINTS MORMON"
replace mormonoth = "Not Mormon" if mormon=="" & religoth!= ""
```

10.

```
encode mormonoth, gen(nmormonoth)
```

Chapter 4 Exercises – Solutions

1.

```
tab faith1

(faith1_w3) F:1. How |
        important or |
```

```
      unimportant is |
  religious faith in |
        shaping how y |      Freq.     Percent        Cum.
---------------------+-----------------------------------
  Extremely important |        472       18.68       18.68
                 Very |        605       23.94       42.62
             Somewhat |        744       29.44       72.06
             Not very |        370       14.64       86.70
 Not important at all |        336       13.30      100.00
---------------------+-----------------------------------
                Total |      2,527      100.00
```

2.

```
tab faith1, sort mis

(faith1_w3) F:1. How |
        important or |
      unimportant is |
  religious faith in |
        shaping how y |      Freq.     Percent        Cum.
---------------------+-----------------------------------
             Somewhat |        744       29.38       29.38
                 Very |        605       23.89       53.28
  Extremely important |        472       18.64       71.92
             Not very |        370       14.61       86.53
 Not important at all |        336       13.27       99.80
                   .d |          4        0.16       99.96
                   .r |          1        0.04      100.00
---------------------+-----------------------------------
                Total |      2,532      100.00
```

3.

```
tab1 faith1 crelder

-> tabulation of faith1

(faith1_w3) F:1. How |
        important or |
      unimportant is |
  religious faith in |
        shaping how y |      Freq.     Percent        Cum.
---------------------+-----------------------------------
  Extremely important |        472       18.68       18.68
                 Very |        605       23.94       42.62
             Somewhat |        744       29.44       72.06
             Not very |        370       14.64       86.70
 Not important at all |        336       13.30      100.00
---------------------+-----------------------------------
                Total |      2,527      100.00
```

```
-> tabulation of crelder

(crelder_w3) R:28. |
   How much do you |
personally care or |
  not about [INSERT |
         LIST A-C] |      Freq.     Percent        Cum.
-------------------+-----------------------------------
         Very much |      1,337       53.06       53.06
          Somewhat |        923       36.63       89.68
          A little |        187        7.42       97.10
Do not really care |         73        2.90      100.00
-------------------+-----------------------------------
             Total |      2,520      100.00
```

4.

```
tab faith1, nol

 (faith1_w3) |
    F:1. How |
   important |
          or |
 unimportant |
          is |
   religious |
    faith in |
  shaping how |
           y |      Freq.     Percent        Cum.
------------+-----------------------------------
           1 |        472       18.68       18.68
           2 |        605       23.94       42.62
           3 |        744       29.44       72.06
           4 |        370       14.64       86.70
           5 |        336       13.30      100.00
------------+-----------------------------------
       Total |      2,527      100.00

tab crelder if faith1==1 | faith1==2

(crelder_w3) R:28. |
   How much do you |
personally care or |
  not about [INSERT |
         LIST A-C] |      Freq.     Percent        Cum.
-------------------+-----------------------------------
         Very much |        693       64.71       64.71
          Somewhat |        313       29.23       93.93
          A little |         46        4.30       98.23
Do not really care |         19        1.77      100.00
-------------------+-----------------------------------
             Total |      1,071      100.00
```

5.

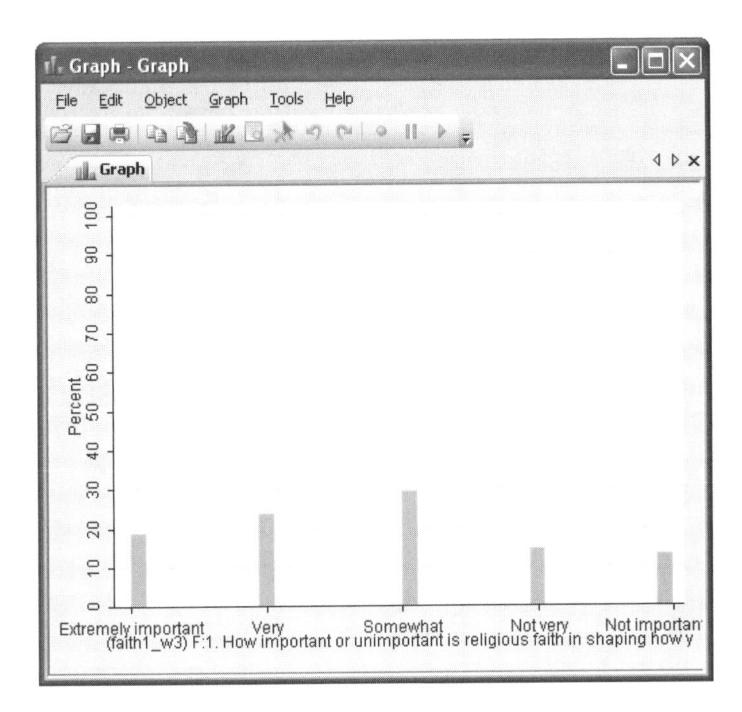
Graph - Graph
File
Edit
Object
Graph
Tools
Help
Graph
Percent
0
10
20
30
40
50
60
70
80
90
100
Extremely important
Very
Somewhat
Not very
Not importan
(faith1_w3) F:1. How important or unimportant is religious faith in shaping how y

6.

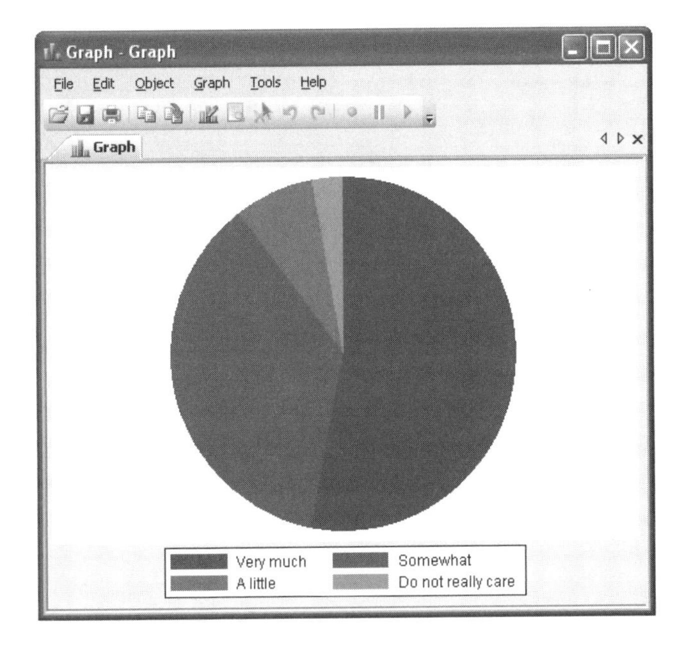
Graph - Graph
File
Edit
Object
Graph
Tools
Help
Graph
Very much
Somewhat
A little
Do not really care

7.

sum kidwntmn, det

```
          (numkidswntmin_w3) R:18. In total, how many
             children do you intend to have? [RECO
-------------------------------------------------------------
      Percentiles      Smallest
 1%            0              0
 5%            1              0
10%            1              0       Obs                2496
25%            2              0       Sum of Wgt.        2496

50%            2                      Mean           2.489984
                        Largest       Std. Dev.      1.348308
75%            3             12
90%            4             13       Variance       1.817936
95%            4             16       Skewness       2.517685
99%            7             18       Kurtosis       21.29259
```

8.

tabstat kidwntmn relretrt, stat(mean p50 sd var) col(stat)

```
    variable |        mean       p50        sd  variance
-------------+----------------------------------------
    kidwntmn |    2.489984         2  1.348308  1.817936
    relretrt |    .7450278         0  2.790436  7.786535
------------------------------------------------------
```

9.

graph box kidwntmn

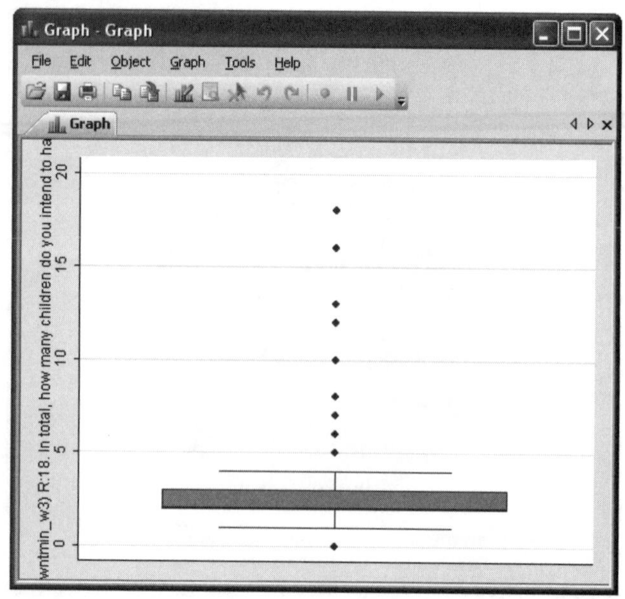

10.

```
hist relretrt, freq
```

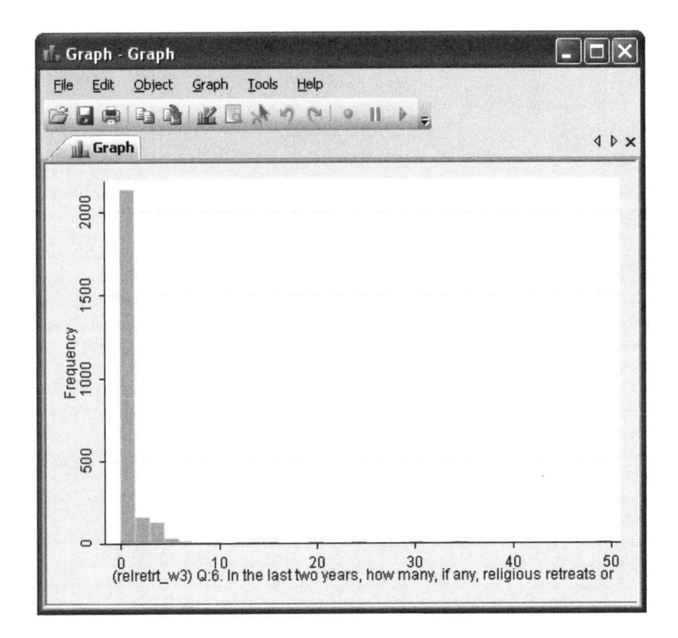

Chapter 5 Exercises – Solutions

1.

```
tab crelder faith1, col

+-------------------+
| Key               |
|-------------------|
|     frequency     |
| column percentage |
+-------------------+

 (crelder_w3) R:28. |
    How much do you |
 personally care or |     (faith1_w3) F:1. How important or unimportant is
  not about [INSERT |              religious faith in shaping how y
         LIST A-C]  | Extremely     Very  Somewhat  Not very  Not impor |    Total
--------------------+-------------------------------------------------------+---------
          Very much |       337      356       339       163        139 |    1,334
                    |     72.01    59.04     45.87     44.05      41.49 |    53.04
--------------------+-------------------------------------------------------+---------
           Somewhat |       110      203       320       160        129 |      922
                    |     23.50    33.67     43.30     43.24      38.51 |    36.66
--------------------+-------------------------------------------------------+---------
```

```
        A little |        12        34        57        36        48 |       187
                 |      2.56      5.64      7.71      9.73     14.33 |      7.44
-----------------+-----------------------------------------------------+----------
Do not really care |       9        10        23        11        19 |        72
                 |      1.92      1.66      3.11      2.97      5.67 |      2.86
-----------------+-----------------------------------------------------+----------
           Total |       468       603       739       370       335 |     2,515
                 |    100.00    100.00    100.00    100.00    100.00 |    100.00
```

2.

```
tab crelder faith1, col chi
```

[TABLE OMITTED]

```
          Pearson chi2(12) = 149.7612   Pr = 0.000
```

3.

```
tab crelder faith1, col chi gamma taub
```

[TABLE OMITTED]

```
                         gamma = 0.2835  ASE = 0.024
             Kendall's tau-b = 0.1924  ASE = 0.017
```

4.

4a.

```
tab attend
tab attend, nol
recode attend (0/1=0) (2/6=1), gen(freqatt)
```

4b.

```
bysort freqatt: tab crelder faith1, col chi gamma taub
```

```
-------------------------------------------------------------------------------
-> freqatt = 0

+-------------------+
| Key               |
|-------------------|
|     frequency     |
| column percentage |
+-------------------+

 (crelder_w3) R:28. |
    How much do you |
 personally care or |    (faith1_w3) F:1. How important or unimportant is
   not about [INSERT           religious faith in shaping how y
```

```
          LIST A-C] | Extremely      Very  Somewhat  Not very  Not impor |     Total
-------------------+-------------------------------------------------------+----------
          Very much |        64       117       220       150       136 |       687
                    |     73.56     59.69     47.11     45.05     41.46 |     48.69
-------------------+-------------------------------------------------------+----------
           Somewhat |        16        60       195       140       125 |       536
                    |     18.39     30.61     41.76     42.04     38.11 |     37.99
-------------------+-------------------------------------------------------+----------
           A little |         3        17        37        34        48 |       139
                    |      3.45      8.67      7.92     10.21     14.63 |      9.85
-------------------+-------------------------------------------------------+----------
 Do not really care |         4         2        15         9        19 |        49
                    |      4.60      1.02      3.21      2.70      5.79 |      3.47
-------------------+-------------------------------------------------------+----------
              Total |        87       196       467       333       328 |     1,411
                    |    100.00    100.00    100.00    100.00    100.00 |    100.00


          Pearson chi2(12) =   58.1602   Pr = 0.000
                     gamma =    0.1986  ASE = 0.034
           Kendall's tau-b =    0.1349  ASE = 0.023

-------------------------------------------------------------------------------
-> freqatt = 1

+-------------------+
| Key               |
|-------------------|
|     frequency     |
| column percentage |
+-------------------+

 (crelder_w3) R:28. |
     How much do you |
  personally care or |     (faith1_w3) F:1. How important or unimportant is
    not about [INSERT              religious faith in shaping how y
          LIST A-C] | Extremely      Very  Somewhat  Not very  Not impor |     Total
-------------------+-------------------------------------------------------+----------
          Very much |       273       238       118        13         3 |       645
                    |     71.65     58.62     43.54     35.14     42.86 |     58.53
-------------------+-------------------------------------------------------+----------
           Somewhat |        94       143       125        20         4 |       386
                    |     24.67     35.22     46.13     54.05     57.14 |     35.03
-------------------+-------------------------------------------------------+----------
           A little |         9        17        20         2         0 |        48
                    |      2.36      4.19      7.38      5.41      0.00 |      4.36
-------------------+-------------------------------------------------------+----------
 Do not really care |         5         8         8         2         0 |        23
                    |      1.31      1.97      2.95      5.41      0.00 |      2.09
-------------------+-------------------------------------------------------+----------
              Total |       381       406       271        37         7 |     1,102
                    |    100.00    100.00    100.00    100.00    100.00 |    100.00

          Pearson chi2(12) =  65.1490   Pr = 0.000
                     gamma =   0.3461  ASE = 0.041
           Kendall's tau-b =   0.2142  ASE = 0.026
```

Chapter 6 Exercises – Solutions

1.

ci longstr

```
    Variable |  Obs   Mean   Std. Err.  [95% Conf.  Interval]
-------------+---------------------------------------------
     longstr | 2211    748          12         724        771
```

2.

ci longstr, level(99)

```
    Variable |  Obs   Mean   Std. Err.  [99% Conf.  Interval]
-------------+---------------------------------------------
     longstr | 2211    748          12         717        778
```

3.

ttest longstr==365

```
One-sample t test
------------------------------------------------------------------------------
Variable |     Obs        Mean    Std. Err.   Std. Dev.   [95% Conf. Interval]
---------+--------------------------------------------------------------------
 longstr |    2211    747.6278    11.87787    558.5124    724.3348    770.9207
------------------------------------------------------------------------------
    mean = mean(longstr)                                          t =  32.2135
Ho: mean = 365                                   degrees of freedom =     2210

   Ha: mean < 365               Ha: mean != 365               Ha: mean > 365
 Pr(T < t) = 1.0000         Pr(|T| > |t|) = 0.0000         Pr(T > t) = 0.0000
```

4.

ttest longstr, by(cu_cohab)

```
Two-sample t test with equal variances
------------------------------------------------------------------------------
   Group |     Obs        Mean    Std. Err.   Std. Dev.   [95% Conf. Interval]
---------+--------------------------------------------------------------------
      No |    1595    630.6307    12.35235    493.3212    606.4022    654.8593
     Yes |     616    1050.567    24.26674    602.2847    1002.911    1098.222
---------+--------------------------------------------------------------------
combined |    2211    747.6278    11.87787    558.5124    724.3348    770.9207
---------+--------------------------------------------------------------------
    diff |           -419.9358    24.94891               -468.8616   -371.0101
------------------------------------------------------------------------------
    diff = mean(No) - mean(Yes)                                   t = -16.8318
Ho: diff = 0                                     degrees of freedom =     2209

    Ha: diff < 0                 Ha: diff != 0                 Ha: diff > 0
 Pr(T < t) = 0.0000         Pr(|T| > |t|) = 0.0000          Pr(T > t) = 1.0000
```

5.

```
anova longstr employst
```

```
                    Number of obs =     2211     R-squared     =  0.0223
                    Root MSE      = 552.887      Adj R-squared =  0.0200

      Source |  Partial SS    df       MS           F     Prob > F
  -----------+----------------------------------------------------
       Model |  15344694.4     5  3068938.88      10.04     0.0000
             |
    employst |  15344694.4     5  3068938.88      10.04     0.0000
             |
    Residual |   674034008  2205  305684.357
  -----------+----------------------------------------------------
       Total |   689378703  2210  311936.065
```

Chapter 7 Exercises – Solutions

1.

```
scatter kidwntmn relretrt
```

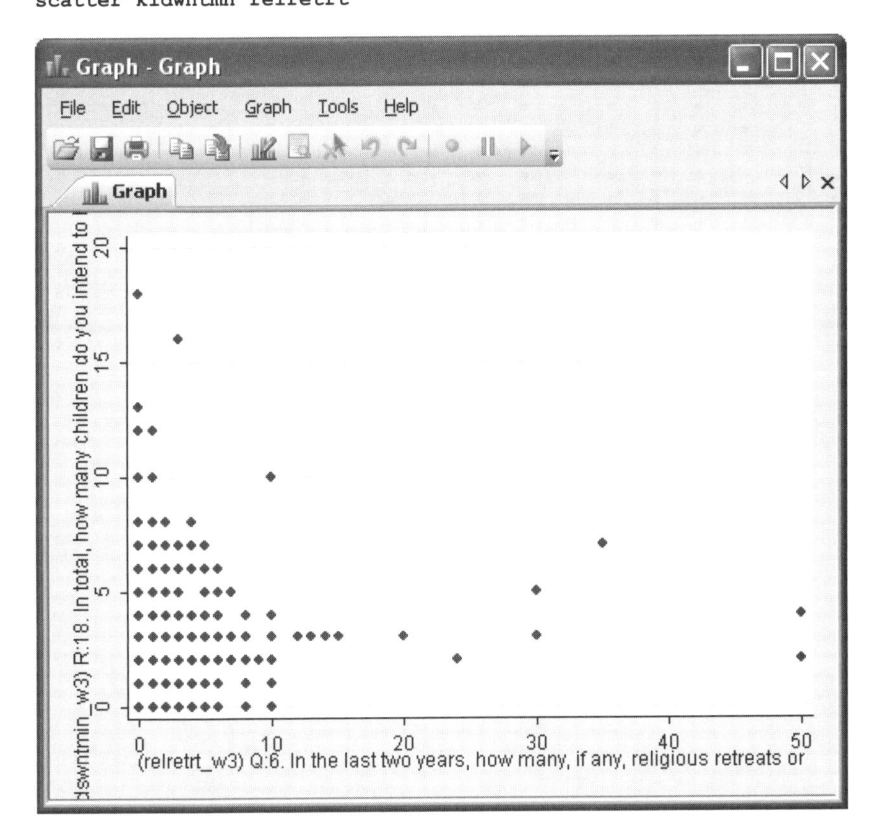

2.

```
scatter kidwntmn relretrt if relretrt<20
```

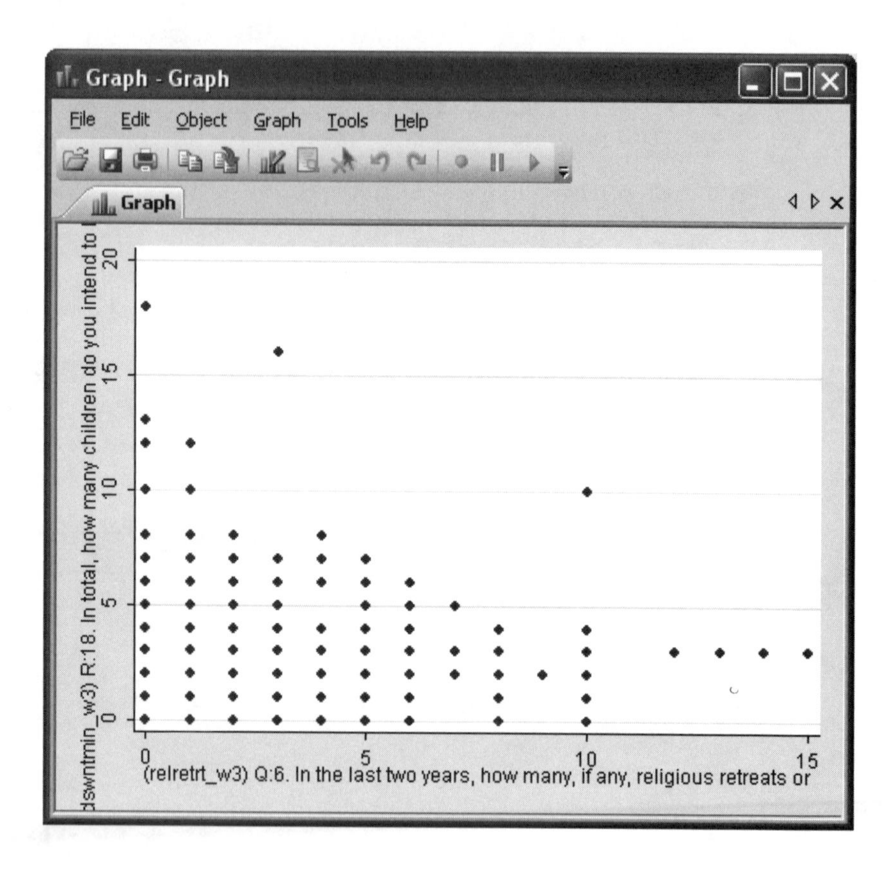

3.

```
corr kidwntmn relretrt
(obs=2488)

             | kidwntmn relretrt
-------------+------------------
    kidwntmn |   1.0000
    relretrt |   0.0865   1.0000
```

4.

```
pwcorr kidwntmn relretrt marrymin, obs

             | kidwntmn relretrt marrymin
-------------+---------------------------
    kidwntmn |   1.0000
             |     2496
             |
    relretrt |   0.0865   1.0000
             |     2488     2514
             |
    marrymin |  -0.1394  -0.0799   1.0000
             |     2454     2463     2471
```

5.

```
reg kidwntmn relretrt

      Source |       SS       df       MS              Number of obs =    2488
-------------+------------------------------           F(  1,  2486) =   18.76
       Model |  33.901597     1   33.901597            Prob > F      =  0.0000
    Residual | 4491.93763  2486  1.80689366            R-squared     =  0.0075
-------------+------------------------------           Adj R-squared =  0.0071
       Total | 4525.83923  2487  1.81979864            Root MSE      =  1.3442

------------------------------------------------------------------------------
    kidwntmn |      Coef.   Std. Err.      t    P>|t|     [95% Conf. Interval]
-------------+----------------------------------------------------------------
    relretrt |   .0416766   .0096216     4.33   0.000     .0228094    .0605439
       _cons |   2.460821   .0278913    88.23   0.000     2.406129    2.515514
------------------------------------------------------------------------------
```

6.

```
reg kidwntmn relretrt marrymin

      Source |       SS       df       MS              Number of obs =    2446
-------------+------------------------------           F(  2,  2443) =   32.56
       Model | 108.495437     2  54.2477184            Prob > F      =  0.0000
    Residual | 4070.76908  2443  1.66629925            R-squared     =  0.0260
-------------+------------------------------           Adj R-squared =  0.0252
       Total | 4179.26451  2445  1.70931064            Root MSE      =  1.2909

------------------------------------------------------------------------------
    kidwntmn |      Coef.   Std. Err.      t    P>|t|     [95% Conf. Interval]
-------------+----------------------------------------------------------------
    relretrt |   .0378432   .0092972     4.07   0.000     .0196119    .0560745
    marrymin |  -.0534883   .0080751    -6.62   0.000     -.069323   -.0376536
       _cons |   3.828593   .2086508    18.35   0.000     3.419442    4.237744
------------------------------------------------------------------------------
```

7.

reg kidwntmn relretrt marrymin gender, beta

```
      Source |       SS       df       MS              Number of obs =    2446
-------------+------------------------------           F(  3,  2442) =   22.23
       Model |  111.092147     3  37.0307157           Prob > F      =  0.0000
    Residual |  4068.17237  2442  1.66591825           R-squared     =  0.0266
-------------+------------------------------           Adj R-squared =  0.0254
       Total |  4179.26451  2445  1.70931064           Root MSE      =  1.2907

------------------------------------------------------------------------------
    kidwntmn |      Coef.   Std. Err.      t    P>|t|                     Beta
-------------+----------------------------------------------------------------
    relretrt |   .0380128   .0092972     4.09   0.000                 .0818961
    marrymin |  -.0524502   .0081169    -6.46   0.000                 -.130103
      gender |   .0655347   .0524912     1.25   0.212                 .0250589
       _cons |   3.768288   .2141455    17.60   0.000                        .
------------------------------------------------------------------------------
```

Chapter 8 Exercises – Solutions

1.

tab godview, gen(dgod)

```
 (godview_w3) [IF BELIEVES IN GOD OR IS |
    UNSURE/DK OR REF] I:5. Which of the |
                                 follow |      Freq.     Percent        Cum.
----------------------------------------+-----------------------------------
God is a personal being involved in the |      1,569       67.43       67.43
God created the world, but is not invol |        246       10.57       78.00
God is not personal, but something like |        477       20.50       98.50
                    None of these views |         35        1.50      100.00
----------------------------------------+-----------------------------------
                                  Total |      2,327      100.00
```

2.

tab1 dgod*

```
-> tabulation of dgod1

godview==Go |
   d is a |
  personal |
     being |
involved in |
  the lives |
  of people |
      today |      Freq.     Percent        Cum.
------------+-----------------------------------
          0 |        758       32.57       32.57
          1 |      1,569       67.43      100.00
------------+-----------------------------------
      Total |      2,327      100.00
```

```
-> tabulation of dgod2

godview==Go |
  d created |
 the world, |
 but is not |
involved in |
  the world |
        now |      Freq.     Percent        Cum.
------------+-----------------------------------
          0 |      2,081       89.43       89.43
          1 |        246       10.57      100.00
------------+-----------------------------------
      Total |      2,327      100.00


-> tabulation of dgod3

godview==Go |
   d is not |
  personal, |
        but |
  something |
     like a |
cosmic life |
      force |      Freq.     Percent        Cum.
------------+-----------------------------------
          0 |      1,850       79.50       79.50
          1 |        477       20.50      100.00
------------+-----------------------------------
      Total |      2,327      100.00


-> tabulation of dgod4

godview==No |
ne of these |
      views |      Freq.     Percent        Cum.
------------+-----------------------------------
          0 |      2,292       98.50       98.50
          1 |         35        1.50      100.00
------------+-----------------------------------
      Total |      2,327      100.00
```

3.

```
help egen
```

4.

```
egen minties = rowmin(datnum numfrien)

tab minties

    minties |      Freq.     Percent        Cum.
------------+-----------------------------------
          0 |         11        0.44        0.44
          1 |        192        7.60        8.04
          2 |        385       15.25       23.29
          3 |        711       28.16       51.45
          4 |        528       20.91       72.36
          5 |        695       27.52       99.88
          6 |          1        0.04       99.92
         20 |          1        0.04       99.96
         25 |          1        0.04      100.00
------------+-----------------------------------
      Total |      2,525      100.00
```

5.

```
mark nomiss
markout nomiss workhrs1 datnum numfrien

sum workhrs1 datnum numfrien if nomiss==1

    Variable |       Obs        Mean    Std. Dev.       Min        Max
-------------+--------------------------------------------------------
    workhrs1 |      2326    21.00903    19.09184          0        100
      datnum |      2326    6.908426    8.719068          0        100
    numfrien |      2326     4.11135    1.066833          0          5
```

6.

```
alpha wommar mandecid wrkngmom

Test scale = mean(unstandardized items)
Reversed item:  mandecid

Average interitem covariance:     .2171798
Number of items in the scale:            3
Scale reliability coefficient:      0.4492
```

7.

```
alpha wommar mandecid wrkngmom, gen(tradgend)

[TABLE OMITTED]

sum tradgend

    Variable |       Obs        Mean    Std. Dev.       Min        Max
-------------+--------------------------------------------------------
    tradgend |      2521    .1243554    .6991233         -1          3
```

"How To" Index

About the Author

Kyle C. Longest received his bachelor's in sociology from Indiana University, and his master's and PhD in sociology from the University of North Carolina at Chapel Hill. Currently, Longest is Assistant Professor in the department of sociology at Furman University, where he teaches Analysis of Social Data, Methods of Social Research, Deviance and Social Control, and Introduction to Sociology. Longest has written articles for a variety of publications, including *Stata Journal*, *Journal of Drug Issues* and *Journal of Marriage and Family*.

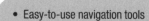